HORSEKEEPING
ON A SMALL ACREAGE

• • • • • • • • •

FACILITIES DESIGN AND MANAGEMENT

BY CHERRY HILL

Illustrations by Richard Klimesh
Photographs by Cherry Hill
(unless otherwise noted)

STOREY
BOOKS

Schoolhouse Road
Pownal, Vermont 05261

*The mission of Storey Communications is to serve our customers
by publishing practical information that encourages personal independence
in harmony with the environment.*

Cover and book design by Judy Eliason
Edited by Deborah Burns
Front cover photograph by Susan Sexton
Back cover photograph by Richard Klimesh
Interior photographs by Cherry Hill, unless otherwise noted
Illustrations by Richard Klimesh, unless otherwise noted

Storey Books are available for special premium and promotional uses and for customized editions. For further information, please call Storey's Custom Publishing Department at 1-800-793-9396.

Printed in the United States by Courier
30 29 28 27 26 25 24 23 22

Library of Congress Cataloging-in-Publication Data

Hill, Cherry, 1947-
    Horsekeeping on a small acreage : facilities design and management / by Cherry Hill ; illustrations by Richard Klimesh. — 1st ed.
        p. cm.
    ISBN 0-88266-597-9 — ISBN 0-88266-596-0 (pbk.)
    1. Horses. 2. Horses—Housing. 3. Farms, Small. I. Title.
SF285.H55 1990
636.1 ' 083—dc20   89-46020
                                                  CIP

Dedication

To Richard for making my horsekeeping dreams come true.
And to Pat Storer for her enthusiasm and love for animals
and her desire to share with others.

Contents

Preface

....

IT WAS A HOT, STICKY JULY DAY IN NORTHEAST Iowa. The auctioneer had moved through most of the household goods and furniture and all of the shop and farm tools. The crowd was thinning and those that were left were congregating under the huge shade trees on the side lawn. My husband and I had purchased a rake and a stepladder and stood leaning against them while we waited for the final item to be sold — the house and ten acres.

The house was definitely modest and the well might have been an early experiment in hand-dug wells. The garage was Model T size but at least there was no old barn that had to be torn down. The ten acres were as flat as a pancake and covered in shoulder-high ironweed. There wasn't a single fence post or rail in sight but there was rich Midwest dirt and sub-irrigated fields. The road out front was lazy and peaceful yet it was only ten miles to a fair-sized town.

Far from ideal, yet if it went for an affordable price, it might make a nice little horse farm. Finally just a dozen folks were left and only two parties were bidding. When the gavel rang for the last time, my husband looked at me with a smile and said, "I guess

I'll put the ladder and the rake back in the garage!" We were ready, once again, to set up horsekeeping!

I simply cannot imagine life without horses. It's not that horses are more important than or a substitute for people or other activities, but horses do have a special way of making life's big picture complete. When my schedule or the weather does not allow me to ride, the day feels as if it has a piece missing. But then riding is just one part of the horse experience. Conscientiously caring for animals brings a wonderful sense of satisfaction. There is nothing quite as fulfilling as a job well done and the pride of owning a healthy, fit, well-trained, and happy horse is great. Just imagine being able to see the fruits of your efforts as you glance out the window to check on your broodmare and foal, as you stroll through your well-manicured pasture, or as you open the door to your tidy tack room to prepare for a morning ride.

Even though owning a horse and boarding it away from home is better than not having a horse at all, keeping a horse at home offers many advantages. It allows you to be involved in and attend to the many details of horse care. And since keeping a horse at home makes it so much more convenient for routine handling and training, you will find you are able to spend much more time with your horses. There is an observation made by country folks regarding the activities of newcomers to the area. The progress of property improvements are noted. If the animal facilities show a marked priority over the home, the person is a true farmer or rancher at heart!

The suggestions and information offered in this book are based on recommendations from extension agents all over North America as well as on my experiences owning and managing horses in Alaska, Arizona, Colorado, Idaho, Illinois, Iowa, Michigan, and Alberta. The herds have ranged from two to more than one hundred horses; from newborns to geriatrics including "idle" horses as well as those in all phases of development and training. I have been involved with breeding operations and training businesses involving both English and Western riding. The size of the facilities varied from one to 160 acres. Some farms were relatively complete on my arrival, others required remodeling or repair, and some were mere tracts of bare land. From these experiences, I would like to share with you what I feel contributes to a horse's well-being.

I hope that this book will not only prove to be a valuable reference for you but will stimulate you to form your own ideas as well. I have approached the subject of horsekeeping by providing information on the behavior and needs of horses before outlining the design of the various facilities and the development of your management scheme. The more thoroughly you understand horses, the more appropriate your plans will be and the more successful your horsekeeping venture. Since the emphasis of this book is facilities and management, I mention health and nutrition topics only briefly. Refer to the Glossary and Bibliography for more information.

My suggestions regarding facilities are not meant to be the final word in such things as horse barns and fences. There are simply too many options to ever discuss them all and the choices are constantly changing. Similarly, by using photos of various products, I do not infer that they are in any way better than other choices. I have included a product reference section in the Appendix and encourage you to use it to start your own research.

By covering various time-honored methods and promising new options in horse facilities and management, I hope that you begin to develop a horsekeeper's consciousness. As you read this book and observe existing horse farms, you will begin to formulate your own mental checklists: questions to ask and characteristics to look for when planning the various aspects of your farm.

Expensive facilities do not guarantee good horsekeeping and, conversely, simple facilities do not, by themselves, indicate poor care. The well-thought-out and conscientiously applied management plan is the tie that binds a venture together and makes it successful. Good management requires knowledge, dedication, and a sincere interest in the well-being of horses.

Acknowledgments

Thank you to the following people for help with
the preparation of the manuscript:

Lynn K. Brown, *Bits and Horses Tax Service*
Bill Culbertson, *retired Colorado Extension Horse Specialist*
Richard Klimesh, *blacksmith and farrier*
Ron Lonneman, *owner Ron's Equipment Company*
Dr. Ginger Rich, *Colorado Extension Horse Specialist*

Thank you to the following manufacturers for
providing photos:

All Weather Fence, Inc.
Barnmaster, Inc.
BMB Animal Apparel Manufacturers
Cherry Mountain Forge
DT Industries, Inc.
Diversified Concrete Products
Jacobsen Textron
Keystone Steel and Wire Co.
Spalding Laboratories
Triple Crown Fence
Vetline

Thank you to the following persons for allowing me to
take photos of their products or facilities:

Sue and Roland Dixon
Colorado Kiwi Latch Co.
Pat and Norm Storer
Connie and Tom McChesney

RaeAnn and Jerry Curtis
Lynn and Ted Brown
Dale and Noel Bormann

SECTION ONE
KNOWING
HORSES

....

BUD WAS A NICE ENOUGH GUY. YOU COULDN'T help but like him—the outdoor type, building a cottage on his acreage on the weekends. I'd see him out there framing the walls, working on the roof. He'd wave as I rode by. My husband and I even invited him over for a barbecue one weekend and thought he would make a nice addition to the neighborhood.

Then, however, he got into horses. A group of horses appeared on his land early one spring. I think he found a package deal and decided horses would be a way to "fit in" to his new "horse neighborhood." Unfortunately, there simply isn't a satisfactory way to care for horses properly on a weekend-only basis. You can't put out three or four bales of hay on Monday and expect horses to ration it so there's some left for Friday. And what if a horse gets hurt on Monday and isn't discovered until Friday evening?

The first sign of Bud's inexperience was the number of horses he put on his ten-acre piece. Geldings and mares of all ages, and two yearling colts that needed to become geldings yesterday! Even with ten acres of space, there were squabbles and minor injuries. And with so many horses, it wasn't long before the grass was reduced to roots, struggling to survive. The acreages on either side of Bud's place were properly managed and still had lush growth, but Bud's land, in contrast, was severely overgrazed. Just before the pasture turned to dirt, however, a stack of hay appeared. He must have received some friendly advice. Things might work out all right after all, if he kept the horses off the pasture while it revived—which might take as long as eighteen months.

The next Saturday morning, however, the veterinarian was called to Bud's place. Later in the afternoon, a neighbor with a back-hoe dug a huge hole toward the back of the pasture. A hole big enough to bury a horse. Apparently, one of Bud's mares had broken a leg early in the week. By the time he discovered it Friday night, she was close to death from infection. The veterinarian said there was no reasonable course of action other than euthanasia.

If Bud had become informed and acted responsibly *before* he became a horse owner, his land would not have been ruined and a horse would not have had to suffer an agonizing death.

CHAPTER ONE

The Benefits and Responsibilities of Horse Ownership

DID YOU EVER WONDER WHY SO MANY TEN-YEAR-old girls collect horse statues and decorate their bedrooms with wall-to-wall photos of horses? Why some, once they have had the opportunity to stroke a real, live horse, are reluctant to wash their hands so as to preserve the heavenly horse smell for as long as possible?

How about the well-educated, capable adult who secures a position as a working student, agreeing to clean horses, tack, and stalls in exchange for riding lessons? Why does the technician or corporate manager feel that the enjoyment he or she gets from weekend rides makes it possible to get through another hectic week?

What makes a horse owner postpone his own medical checkup but religiously schedule routine appointments with his horse's veterinarian? Why does a person consistently skip his breakfast or vitamins ("I'll catch up later") and grab fast food ("Just this once") yet never dream of using short-cuts in his horse's rations? What makes a horse owner forgo a new coat yet not bat an eye when slapping down several hundred dollars for a new winter horse blanket?

The reasons behind these and other "horse-crazy" behaviors are largely due, in my estimation, to the noble values that horses represent to humans. As it has often been said, there is something about the outside of a horse that is good for the inside of a man. When we treat horses with the respect they deserve, they provide us with many unique opportunities to find a type of nobility in ourselves as well.

It has frequently been noted that animals are a reflection of their owners, and in no case is this more evident than with the home-raised, home-trained, home-kept horse. A healthy relationship between a human and horse is a partnership. Both have certain obligations to each other, and when those are met consistently on both sides there is the foundation for a happy and productive exchange.

Responsibilities

Although there is something almost magical about working with horses, they are often over-romanticized in stories and movies and the realities involved in horse training and

SAMPLE BUDGET PER HORSE PER YEAR

FEED:

Hay *(20 lbs. per day x 365 days = 7300 lbs.*
or 3.65 tons x $120 per ton) $438.00

Grain *(4 lbs. per day x 365 days = 1460 lbs.*
x .10 per lb.) 146.00

BEDDING 150.00

VETERINARY SUPPLIES AND CARE:

Immunization 30.00

Deworming *(6 times per year x $9)* 54.00

Dental and misc. 50.00

Farm call charges 75.00

FARRIER:

Shoeing *(6 times at $50)* 300.00

Trimming *(3 times at $14)* 42.00

TOTAL **$1,285.00**

care are glossed over. Horse ownership requires a substantial investment of money, time, hard work, and sincere dedication.

The domestic horse depends on you, since he is prevented from taking care of himself. Horses need care when they are idle as well as when they are being actively ridden. Their requirements do not diminish if their owner's interest does. As you will see in a later chapter, during the winter when you are least likely to ride your horse regularly, he actually needs the most care. Following are some of the realities of horse ownership:

Money. The initial purchase price of a horse is just the beginning of your costs. You can certainly cut costs by being innovative, but horsekeeping does require weekly expenditures.

Time. You must be willing and able to spend time attending to your horse's needs every day at least two times per day. You will have to tend to feeding, cleaning, grooming, and exercise every day, as well as associated chores such as buying feed, repairing tack and facilities, and more. See chapter 3 on horse needs for more details.

Hard work. Many parts of horse ownership involve hard physical labor, not only the energy-expending kind but the "back-breaking" kind as well! Shoveling manure, toting bales, carrying water, giving a vigorous grooming, and instituting a conscientious exercise program for your horse all go more smoothly if you are physically fit.

Trade-offs. When you own horses, there will sometimes be occasions when you must give up other things that you like to do—like sleep, warmth, and comfort—to assure that your horse receives proper care. Horses do pick inconvenient times to have foals, become ill, or get injured: the middle of the night; just as you are leaving for an important meeting in your three-piece suit; during the worst blizzard your area has seen in over fifteen years; or moments before the kick-off of the championship football game. Even routine horse care will sometimes seem to intrude on your other plans: your veterinarian can make it only on the morning you planned to get together with a special friend; the day before you plan to leave on vacation, the person you have lined up to do chores becomes unavailable; and so on.

Legal. Horse owners have legal obligations to their horses, to neighbors, to other horse owners in the area, and to pedestrians and motorists passing by the property. Check the liability laws that apply to your specific location. In terms of horse care, your legal obligation may be described by a phrase such as "ordinary care and diligence," which can be open to a wide range of interpretations.

The Benefits of Horse Ownership

The relationship between a man or a woman and a horse is often very fulfilling, without the complications of human associations. A horse doesn't talk back but does tell you, using body language and other non-verbal communication, how it interprets your actions. A horse will reveal your true character—the quickness of your temper, how consistent you are—and can give you the chance to become a better person. Caring for and interacting with horses can make a person more reliable, thorough, trustworthy, honest, and consistent. People who have difficulty working with other persons often find that a horse can teach them the meaning of teamwork.

Once the framework of a healthy partnership has been firmly established, a horse can also become an affectionate companion. It is dangerous to encourage this part of the relationship to develop before the business agreements have been made, however, as a horse is simply too big to be a cuddly lap dog and it must know the acceptable limits of interaction with humans. Yet there is no reason why a horse cannot be a partner and a friend as well. There is nothing quite as uplifting as a soft nicker greeting you at the gate when it is time to go for a ride. You may have heard the saying, "There is no secret so great as that between a rider and his horse."

A trustworthy horse can provide invaluable therapy for people caught up in a hectic pace. Riding can help a person shed stress and stop the mental conversations that cause it. Few experiences equal a trail ride in the fresh air, especially if there is gorgeous scenery. Riding down a road or in an arena, however, can also be enjoyable and beneficial for both the horse and rider in many ways. There is nothing quite like a rein-swinging walk to lull a person back into natural rhythms; nothing like a brisk trot with its

A horse is a feast for the eyes and provides an opportunity to observe animal behavior. COURTESY OF PAT STORER

metronome-like quality to invigorate one physically; nothing like a rollicking canter cross-country to rekindle the sensations of freedom.

The exercise associated with the care and riding of horses also adds to a person's fitness. Grooming, cleaning, tack, and riding involve many muscle groups and types of activities; the composite exercise is well-balanced and certainly not monotonous.

Another physical benefit is that horses satisfy the human need for contact, the desire to touch and be in close proximity to a warm, responsive being. By this I don't mean to encourage petting the horse; petting is more appropriate for dogs than horses. Beneficial physical contact with horses occurs during grooming and riding and the occasional hug around the neck. The feel of a soft, sleek coat coupled with the rich, earthy smell that is so characteristically *horse* is very endearing.

And horses are a feast for the eyes. They are beautiful to watch resting, grazing, playing, and moving with energy and grace. They provide a valuable chance to learn about animal behavior. Their reactions and interactions are fascinating and provide material for stories and exchanges.

Keeping horses at home will teach you many interesting things about behavior, breeding, selection and use of tack, various styles of riding, training, health care, and much, much more.

While some people may consider care-giving a responsibility, others see it as an opportunity to nurture. Taking care of a horse's needs can help a person establish good habits and routines and bring an otherwise elusive order to a chaotic life.

Finally, being involved with horses offers social benefits. There are many local, regional, and national organizations that are designed for family participation. Groups are available for all types of horse involvement: trail riding, lessons and clinics, competitions of all levels and types, and groups for "backyard horsemen" of varying interests. Besides providing a great place to share experiences, horse groups are a good place to exchange ideas, form friendships, and create a network for group purchases and business transactions.

CHAPTER TWO

Horse Behavior as it Relates to Management

IN ORDER TO MAKE WISE DECISIONS IN PLANNING your facilities and devising your management plan, you must know how and why horses behave as they do. You are not going to significantly change patterns that have been a part of the horse for over 60 million years. It is best therefore to design facilities that are specifically suited to horses and their behavior patterns.

To understand horses, you must first realize and accept that horses are not humans, puppy dogs, or art objects. Even though horses can elicit emotions similar to those that are felt for family or friends, dealing with horses as though they were humans is a dangerous anthropomorphic trap. You will see specifically what I mean in this chapter.

Horses should not be thought of as pets but as partners. Although they can be cute and even charming, remember that they are large and potentially dangerous animals. Excessive doting can change your relationship from a partnership to one of human subservience. Horses are much more content if early on they are made to know the rules and are consistently and fairly treated according to those rules.

Finally, although horses truly are works of art, whether peacefully grazing or in breathtaking motion, they really aren't collectibles like porcelain statues or framed oils. They are organic creatures with specific behavior patterns and very real needs.

Behavior Patterns

Often horses' actions are interpreted as misbehavior when in fact they are instincts that have been inherited as proper horse conduct. While a horse's natural behavior must be altered somewhat before it can be useful, the trainer must work with already existing instincts and reflexes for minimal stress and long-lasting results. Observations of horses in the herd and in various styles of domestic confinement will give you insight and help you make good handling and management decisions.

First you must realize that whether or not there is action, there can be misbehavior. A sullen horse, rigid and unyielding, is misbehaving just as much as is the wildly bucking one. And any behavior that is

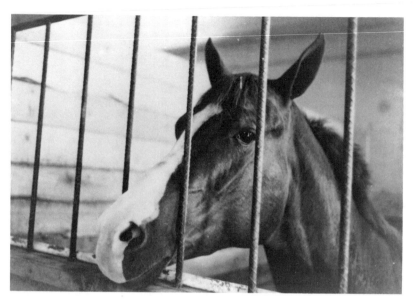

Horses are social creatures; they can get very lonely if isolated from other horses.

and herd companions participating in social rituals—all of these are examples of the horse's social behavior.

Mutual grooming. When two horses have a strong bond, one way they show their mutual admiration is to demonstrate the saying "You scratch my back and I'll scratch yours." You may have seen two horses standing head to tail, nibbling each other along their backs.

Sometimes horses innocently attempt to interact with humans as though they were horses. When you groom a horse the rubbing sensation, especially in the vicinity of the withers, causes him to want to reciprocate by nuzzling you. Even though the gesture is meant to be friendly, not aggressive, intentions don't count. After all, you would not want your horse to massage the back of your neck with his teeth, would you? Such an act must therefore be discouraged. Say "No!" and loudly slap the horse's neck or shoulder with the flat of your hand or a leather strap, and then immediately resume grooming. After two or three attempts at "mutual grooming" with the handler, most horses catch on and see the difference in acceptable behavior between band-mates and humans.

Herd bound. If a horse has not been sufficiently socialized with its handler to the point that it feels secure away from other horses, it may desperately attempt to stay near or communicate with a preferred companion. The classic case is often referred to as "barn-sourness" because the insecure horse links comfort, companionship, and food with the barn. A horse separated from other horses may pace incessantly back and forth along a fence line, paw or weave in its stall, or scream shrilly in an attempt to maintain contact.

What may have originated in the immature horse as a temporary insecurity may evolve into a long-standing and dangerous habit. In order to ensure that such

repeated becomes habit, even though it was not part of a formal lesson. Horses are constantly learning as a result of their handling and the environment.

Even though the modern horse is relatively safe from predators, its long history of struggle for survival has resulted in a deeply imbedded suspicion of anything unfamiliar. Because of this, the modern horse is one of the few domestic animals that still retains the instincts necessary to revert to a wild state. These instincts make the horse awe-inspiring and frustrating at the same time.

The horse is a gregarious nomad with keen senses and instincts and highly developed reflexes. These characteristics are what can send a performance horse to the winners' circle—and a panic-stricken horse through a wire fence. The good horseman must understand the nature of the horse and use it to his advantage.

Gregariousness. Gregarious animals are sociable and prefer to live and move in groups. If the domestic horse has a choice, he will not stand alone but in close proximity to another horse, finding safety and comfort in numbers. An entire band panicking from an imaginary beast, a group huddling tightly against the wind or snow,

N. J. WILEY

The pecking order that exists among horses can lead to fighting at feeding time. Therefore, if you must feed a group, divide the feed into several extra portions and separate the piles with enough distance so that the timid horse can find an unclaimed pile on the fringe.

a bad habit does not get started, horses should be handled separately from a very early age and should never be allowed to panic. The distance and time away from the other horses should be gradually increased. The horse that is quiet and attentive to the handler should be rewarded by occasionally feeding or grooming it away from its companions. It also helps if the horses left in the group (at home) are content with other companions nearby or preoccupied with feed so that they do not call or answer to the horse that is learning to be away.

Pecking order. Just because horses desperately want to be with other members of the band, however, doesn't mean all horses get along well. Especially when there is limited food or space, personality conflicts will surface. Battles may be fought with teeth and hooves or merely with threatening gestures. Once the clash is over, a pecking order or dominance hierarchy emerges. This social rank makes future aggression unnecessary unless a particular horse is not thoroughly convinced of its status and

continually tests the horses immediately above it. The most assertive horse generally earns its choice of feed, water, and personal spaces.

That's why in planning facilities, you should avoid places where horses can get cornered and hurt. You almost have to assume that horses will fight at feeding time, so you should either feed every horse separately or feed groups of horses in a large space and with more hay piles or feeders than there are numbers of horses.

Because of the potentially violent behavior associated with the establishment of status in a pecking order, new horses must be added very carefully to an already established group. It is best if the new horse can spend a few days in close proximity but not in direct contact with the band. However, putting a new horse across an unsafe fence from the group it will later join can result in injury. So, if possible, allow the new horse to interact with a few members of the group across a safe fence. After a few days, put the new horse in with the sub-group. After a few more days, combine all of the horses.

Humans occupy a rung on the ladder of power also, and you will be tested by various horses to see just where you stand. You must convince each horse that you are the top gun. Often a confrontation will take place at feeding time. As you approach a feeder, a horse may come forward aggressively, perhaps with laid-back ears and threatening body language. The worst thing you can do at this point is to reinforce the horse's aggressive behavior by dumping the feed and leaving the pen. You will have rewarded the horse for its pushy behavior.

Instead, the horse must be made to back off and stand at attention until you give a clear signal that it can approach the feeder. In some instances it might be necessary to halter the horse at feeding time and establish control in-hand first. In other cases, you might just have to stomp a foot toward the horse to get its attention. Or you may have to slap it across the chest to keep it from crowding you. You should issue a firm voice command such as a stern "Wait!" and require the horse to remain attentive until you leave the pen. Then use a command such as "OK" to indicate it may now approach the feeder. If a horse continues to intimidate you or make you fearful, seek professional help or get rid of the horse.

Routines. Horses typically perform daily routines in response to various needs. Many of the routines are socially oriented: small groups graze in tight-knit bands on huge ranges, participate in contagious pawing and rolling sessions, or engage in running and bucking games. At regular times of the day, individuals in stalls or groups on pasture can be observed to eat, drink, roll, play, and perform mutual grooming. The desire to participate in these rituals is not diminished, and in fact is probably intensified, for the horse in confinement.

The horse's strong biological clock is noisily evident near feeding time. Once a horse has established a routine of urinating in his stall, he will often, much to the stall cleaner's dismay, "hold it in" all day in the pasture only to flood the stall the instant he is returned to it. In spite of bathing, clipping, and blanketing, most horses love a good roll in the mud, much to the chagrin of the human groom! And the behavior inspiring the old adage, "You can lead a horse to water, but you can't make him drink" is based on firmly implanted habits and governed by a biological clock (and can also be influenced by a horse's keen sense of smell and taste).

Nomadic. The horse is a born wanderer. This nomadic tendency is responsible for confinement behaviors such as pawing, weaving, and pacing. These vices are a response to inactivity, lack of exercise, overfeeding, and insufficient handling. Regular exercise is essential for the horse's physical and mental well-being. Adequate turn-out space and time prevent the development of neuroses. (See the end of this chapter for a more detailed discussion of vices.)

Horses that are kept in box stalls or small pens need to be turned out and allowed to "be horses." Otherwise they may become either very bored with their existence or extremely hyperactive. An introverted horse that has "tuned out" may be lazy, unresponsive, and balky. The overly energetic horse is often anticipatory, nervous, and irritable and can be unsafe.

Senses. Keen senses allow the horse to pick up very slight changes in the environment. More sensitive to subtle movements, far-off sounds, smells, and possibly barometric pressure than are their human handlers, horses are frequently alerted to potential danger while we notice nothing out of the ordinary. Horses are capable of feeling vibrations through their hooves warning them of approaching predators or other horses.

Horses have a very discerning tactile sense. Their lips, skin, and hairs accumulate information which we normally gather with our hands. Horses are dexterous with their lips and can open gates with intricate latches. They can also determine if an electric fence is operating by checking it with the hairs on their lips.

With all of these keen senses ready to put him on red alert, it seems unfortunate that a horse's vision doesn't provide much help in resolving his apprehensions. First of all, a horse has blind spots—the areas directly in front of its face, below its head and neck, and directly behind it. It also has a lesser ability than we do to focus both near and far, which is demonstrated by its wide range of head and neck positions when trying to see something—from craning and straining to lowering and peering.

The horse's eye is slower than ours to adapt to light changes, which explains why a horse must take a few more seconds to get its bearings when stepping out of a dark barn into the bright light or from the bright light into a dark trailer. And finally, because horses see with both monocular and binocular vision at the junction of the two fields of vision, images are thought to jump and be blurred, causing visual distortions and concerns for the horse.

Horses generally have an avid sense of curiosity. They are not content just to look at an object—they must inspect, fiddle, meddle, smell, nuzzle, paw, knock over, and in general dink around with almost anything they can get to.

Flight. When the horse is convinced that danger is imminent, it almost always chooses to flee rather than to fight. It is the rare horse that on its own volition will stick around and reassess the situation in the event it might be imagining things. Horses can be taught to trust their handlers' good sense, however. The horse out on pasture, left to its own devices, would probably avoid the "black hole" that in reality is only an eight-inch-deep spot in the creek. When the trainer (that treats her horse fairly) assures the horse by voice and body language that it is safe to step into the water, the trusting but skeptical animal will reconsider. As long as you make wise decisions and never ask your horse to negotiate something unsafe, your horse's instinctual fears can be overridden by his confidence in you.

Memory. If, on the other hand, a horse lacks confidence or has received poor handling it can behave very unpredictably and spook with the slightest provocation. Because a horse has an excellent memory, it will remember quite remote experiences, especially if they relate to its imagined safety. Horses are believed to never quite forget these fears. All a handler can hope for is to bury the bad experiences with layer upon layer of good ones.

For example, suppose a horse is turned out in a new pasture for the first time. As he trots around snorting, with head high, inspecting the boundaries, a couple of dogs pop out of a wooded area at the edge of the pasture and begin chasing him. In his panic to escape his modern-day predators, he mindlessly heaves his body at the wire fence and manages to stretch and break enough wires to allow him to return to the barn. The stray dogs quit the chase as they are leery of the humans usually around the barn; your horse stands quivering and bleeding alongside one of his buddies.

What do you think will cross your horse's mind the next time you turn him out on that pasture? Even if there are no dogs present, do you think he might avoid the wooded area altogether? Will every moving leaf in the woods make him suspect that killer dogs will emerge? Will he go through the fence again? Unfortunately your horse will be innately suspicious of that pasture, and especially the woods, for a long time. The best plan is to prevent such things from occurring in the first place. Once something traumatic does happen, however, you must allay your horse's apprehension by systematically planning good experiences to bury the bad ones.

A horse who reaches into his water tub

for a drink and receives a shock from a tank heater with an electrical short will very likely refuse to drink even if his body is in a life-threatening state of dehydration.

Reflexes. Horses can assume thundering speeds from a standstill. They can rise from a recumbent sleeping position and instantly run. They can strike or kick in the blink of an eye. These lightning reflexes helped the horse survive for more than sixty million years. The same automatic responses allow today's horse to perform in a vast array of spectacular performance events—but they can also prove to be dangerous for the human handler.

Much of training is designed to work with and/or systematically override the horse's natural reflexes. An example is the withdrawal reflex. This is the natural reaction of a horse to pick up his leg when something touches it. In order to be able to wash, clip, and bandage the legs, you must override this reflex so the horse will keep his hoof on the ground as you touch his leg. But you must keep in mind that you will also want to pick up a horse's hoof to clean it, so you will want to make a discernable difference in the way in which you request each behavior.

Vices

Vices are undesirable habits that horses often develop for legitimate reasons and demonstrate in the stable or in other confinement. Vices, such as cribbing, pawing, and weaving, tend to be performed in the privacy of a horse's stall, often when humans are not around. Bad habits, on the other hand, are undesirable behaviors that horses learn in response to training and are almost always performed during handling or riding. Examples of bad habits are rearing, bolting (running away), and biting the handler. Bad habits can be avoided with proper training. (See Bibliography for relevant books.)

Vices and bad habits are both formed in response to conflict, uncertainty, or restriction. Conflict occurs when a horse has two opposing urges, both equally strong. The thirsty horse who has been shocked by his waterer wants the water but does not want the shock. He may stand near his waterer and paw a hole to China. Unfortunately, even if the waterer is fixed and the horse drinks from it again, he may continue to paw. Vices often remain even though the cause is eliminated. That's why an ounce of prevention is worth a pound of cure.

Uncertainty results when the horse is faced with a problem beyond its power of resolution. Horses do not rank very highly as problem solvers. They would not do well negotiating a maze even if the reward at the end were their favorite food. This is why a horse may panic and pace up and down a fence line at feeding time if the route it must take contains an unusual combination of turns, gates, or doorways. The very calm horse familiar with the routine may easily find its way through the barnyard to its feed, but the high-strung, uninitiated newcomer may race back and forth, looking at the barn, but not know how to get there. If such a behavior were to be repeated several times in a row, it could become an established vice like pacing, which then might occur every time the horse was stressed.

Restriction comes when a horse is limited in his movement. Keeping a horse in a stall or small pen is contrary to its desire to roam and have regular exercise. Prolonged confinement is one of the leading causes of such vices as pawing, pacing, weaving, and stall walking.

Remember, vices can be prevented with proper management. Examples of some common stable vices follow.

Wood chewing. The beaver-like gnawing of wood rails, planks, buildings, and feeders is costly, unnecessary, and can be dangerous to the health of the horse. Wood chewing can afflict a horse of any age and can result in colic from wood ingestion or damage to the gums and lips from splinters, to say nothing of the damage to facilities.

Young horses may begin nibbling out of boredom, curiosity, or perhaps to relieve an itching of the gums during teething. Serious wood chewing can initially be caused by low fiber intake in relation to a horse's needs, especially during cold and/or wet weather. Horses appear to be relaxed and comforted when they are able to spend a good deal of time chewing long hay. Horses deprived of this natural satiation may be seeking oral gratification and an increase in fiber intake from the wood. Weather-related wood chewing is thought to be a result of the frustration and anxiety felt when the animal is uncomfortable. Water also softens the wood, making it more palatable and aromatic.

Be sure your horse's diet is well balanced and adequate in fiber in the form of long-stem hay. Provide the animal with ample exercise. To cure the chronic wood-chewer and to prevent others from acquiring the vice, coat all wooden surfaces with one of the wood preservatives listed in the chapter on fences. Cover all wooden edges in stalls with metal corner trim. Run electric fence wire along the wooden fence rails.

Cribbing. Cribbing, or wind-sucking, is a very debilitating vice. The cribbing horse grabs the edge of a partition, the top of a door or post, a feeder, or another solid object, arches its neck, and swallows air in labored gulps. Although, at first glance, you may think a cribber is a wood-chewer or vice-versa, these are two very different behaviors. The horse that was a wood-chewer first and later a cribber probably would have become a cribber even if he was kept in an all-metal stall. Cribbers are often nervous, neurotic individuals who find comfort in their ritual. Unfortunately, cribbing can result in abnormal wear on the teeth and in colic from swallowed air. It is a difficult, if not impossible, vice to cure. Cribbers are often more interested in crib-

bing than in eating and waste a lot of energy on the vice, so they tend to be "unthrifty" individuals. Research shows that cribbing releases narcotics (opiates) from the brain, giving the animal a natural, habit-forming high. Cribbing can be a contagious social behavior so it is best not to keep a cribber on your property.

Pawing. Pawing is initially a signal that a horse wants or needs something, but once a horse has been allowed to perform such behavior, it may become a habit that no longer has any specific cause. The wild horse or the pastured horse uses pawing for many practical purposes such as uncovering feed under snow, opening up a water hole, digging up roots during a dry season, inspecting an unfamiliar object, and softening the soil before rolling. Pawing can also indicate pain and restlessness as with a colicky horse or a mare that is foaling or expelling a placenta.

N. J. WILEY

Lack of training, lack of sufficient exercise, boredom, and confinement can cause a horse to paw. Pawing is an expression of the horse's desire to wander. Pawing is damaging to the horse and the facilities and must be prevented by proper management and training.

The pawing instincts of the stalled horse are especially evident around feeding time: many horses paw to indicate their grain or water pail is empty, and some just paw in anticipation of being fed. Unfortunately, feeding a pawing horse is a form of reward and encourages him to repeat his behavior in the future.

Other horses paw as a response to domestication pressures. Confinement, lack of exercise, and over-feeding often create a horse with excess energy. As a replacement for exercise, the normally active horse that is confined will paw, in some cases by banging the front feet on the stall door. In other situations, the hoof may not actually contact the ground but just makes repeated swipes through the air.

A good number of horses seem to like the sounds associated with pawing and many seem to derive an extra measure of satisfaction from bearing down and really scooping up the ground. This results in great damage to stall floors, hooves and joints, and shoeing. One of the greatest causes of lost shoes is the habitually pawing horse who either loosens the clinches from the repeated pounding or catches the shoe on something like a fence and pulls the shoe off. Such continual concussion can also lead to joint problems and raises havoc with the anterior-posterior balance of the unshod hoof—the horse is constantly wearing off the toe as the hoof scrapes across the ground.

Horses that have not been made to accept the confinement of cross-ties, a hitching post, or a trailer often paw out of impatience or nervousness. Similarly, the unschooled horse at the beginning of a race or the stallion waiting to service a mare may also paw out of impatience. Insecure horses, when separated from the herd, paw from fear of being alone.

Because pawing is damaging to the horse and to the facilities and can really wear on the nerves of someone working around the barn, the habit should be discouraged early. Approach the problem from a preventative standpoint, using a psychological approach. Be sure that the horse receives conscientious handling and adequate exercise and turn-out and is not being over-fed or inadvertently being rewarded for pawing at

feeding time. These are the best safeguards against the unwanted habit. If the pawing appears to be caused by boredom, a diversionary stall toy may be the answer.

For the horse with a firmly established behavior pattern, it will probably be necessary to eradicate the habit with physical means (see Bibliography). Hobbling is one of the best techniques to prevent pawing. Introduce hobbling as part of a series of restraint lessons, generally after the horse has learned to stand still by means of halter, chain, or hind leg rope. If the hobbles are applied before the horse has developed some respect for restraint, he will probably learn to hop around rather well with the hobbles and this is not desirable. Once the horse accepts hobbles they can be used when the horse is most likely to paw, such as when he is tied and left unattended at a hitching rail.

Hobbles aren't really safe for the stalled horse and it is impractical to try to catch a horse pawing in his stall and discipline him each time, so it is necessary to use a means of self-punishment instead. Success has been obtained by suspending a ten-inch piece of chain from a leather strap that is buckled above the fetlock. You may add a 3-inch cube of wood at the end of the chain. Each time a horse attempts to paw, he instantly receives a punishment and soon is discouraged from the act. The question is raised as to whether this method might damage the horse's fetlock or pastern. Horses usually quit pawing after a few encounters with the chain or block, so they may receive just a few blows. If the horse's habit is ignored, however, he receives a daily quota of concussion and is destructive to facilities as well.

Weaving. The horse who weaves stands with his head and neck over a stall door or fence and sways his body from side to side. The rhythmic, lulling movement appears to be soothing to a nervous horse or to one who has insufficient exercise. It does use a lot of energy, so often a weaver is a horse in poor condition. This obsessive, repetitive movement can wear unshod hooves unevenly and even distort the growth of the hoof wall.

A horse confined an excessive amount of time (for a week or so) may try this behavior to fill his need for exercise. If he is then properly exercised, he may still retain the habit, even out in a pasture. Other horses can mimic this behavior, so it is best to prevent such a vice from becoming established.

Tail (and body) rubbing. Here is a vice that, like weaving, is a rhythmic, swaying motion. Like most vices, it may have begun for a legitimate reason, but it remains long after the cause has been removed. Horses may begin rubbing due to shedding, a lack of grooming coupled with the inability to roll (self-groom), or as a response to parasites, a discharge under tail and subsequent matting, or a dirty anus, udder, or sheath.

The chronic rubber is hard on facilities, actually knocking rails down, stretching wire fences, breaking branches and small trees, and damaging stall walls. The horse who rubs his tail or mane may end up with bald spots.

Stall kicking. Few vices can be as destructive to both the horse and the facilities as stall kicking. In some cases, the horse stands with its hindquarters near a wall and rhythmically thumps the wall with one hind foot while the head bobs in a reciprocating motion. Other horses let fly with both hinds at once. There can be several such explosive bursts in a row, but because of the energy and balance required, this double-barreled kicker cannot produce the characteristic metronome-like thudding of the one-legged kicker. The explosive kicker can wipe out a stall wall in a single kicking bout, to say nothing of the damage that can be done to its hind legs—primarily from the hock down. Capped hocks and curbs are often associated with chronic stall kickers.

Some horses have a predisposition to neurotic breakdown when faced with in-

sufficient exercise, excess feed, or constantly changing neighbors. This tendency may be genetically inherited, it may be formed from early experiences with the dam or with stable management, or it may simply develop later in life. Often, when such a tendency is coupled with a precipitating cause, such as insufficient exercise, the result is a vice such as stall kicking.

For the chronic stall kicker, you might ask your farrier to fix a special shoe to discourage the dangerous habit.

Like many stable vices, stall kicking may be socially contagious. Neighboring horses might be caught up in the rhythm or may interpret the action as threatening. Stall kicking is not always an act of aggression toward another horse, however. It may be a means of reacting to domestication or training stresses or it could be a playful diversion caused by boredom. Kicking is a part of socially acceptable play among horses and may begin as a natural behavior and then get carried to extremes. It is thought that some horses actually like to hear the sound of the thumping, so some success has been achieved by padding stall walls or hooves.

Neighboring horses may not get along for a variety of reasons. A mare that has gone out of heat (and in some instances, those that are in estrus) may kick at the horse in the next stall whether it is a stallion, a gelding, or another mare. Others may have a personality or pecking order conflict that is unsettling. And some horses that are protective of their feed may use kicking as a defensive measure.

Some horses have learned that a great way to elicit attention from owners, and perhaps even get an extra measure of feed, is to kick. The noise brings a human to the stall, often with a diversionary flake of hay. This not only gives the horse what he wants but it actually rewards him for the kicking behavior. With the long-standing confirmed kicker, there may be no cure, but there are better ways to deal with the vice than using feed as a distraction and inadvertent reinforcement.

As with all undesirable habits, treatment can employ psychological or physical means. Any treatment is more successful if applied before the habit becomes deeply ingrained. A psychological cure requires identification of the cause, while a physical cure is based on elimination of the symptom of the problem. If stall kicking is obviously due to boredom or confinement, much success is obtained by giving the horse additional turn-out time, putting him in a pasture situation or run, giving him a stall toy (see chapter 3, page 25), and regulating his feed. And if a horse's kicking is due to a particular neighbor, merely shifting the horse's location in the barn may help.

If, on the other hand, the horse has developed a chronic kicking habit or a non-specific crankiness, it is best to deal with the symptom of the vice—to treat the kicking itself. Cross-hobbling or side-lining a horse would most likely prevent kicking but may be dangerous if the horse is left unattended. Affixing horizontal boards on edge around the inside of the stall might help: at rump height they may prevent the horse from getting close enough to the wall to kick and fixing them at hock height will punish the horse on the back of the leg and prevent him from contacting the wall with his hoof. An old remedy suggests hanging a heavy iron object above the area the horse kicks so that it hits him on the top of the rump as he raises it to kick.

Kicking chains are often successful: a chain is suspended from a leather strap which is fastened above the hock. The horse is reprimanded by the chain each time he kicks. A simpler self-training device is a specially shaped horseshoe. It should be circular and brought together at the heels so that it just fits over the leg at the cannon and will slide down over the fetlock and rest on the coronary band. Every time the horse kicks, the shoe bangs the pastern. The question may be raised whether methods like these are humane. If a horse has developed a habit of relentless pounding which results in damage to the legs and broken and splintered boards, a fairly severe method needs to be used. Before trying the chain or shoe, however, be sure that the horse receives adequate exercise, appropriate feed, and has reasonable neighbors.

CHAPTER THREE

The Needs of a Horse

T O BE HEALTHY AND CONTENT, A HORSE NEEDS adequate feed, water, shelter, exercise, rest, companionship, and veterinary and farrier care.

Feed. Feeding horses is an art and science unto itself, and entire volumes are devoted to the subject. Refer to the Bibliography for books that will tell you how to select feed and balance a ration for horses of all ages and activity levels.

Understanding certain principles ahead of time will help you make appropriate plans for facilities and management.

● Since horses evolved as grazers, their systems are adapted to many small meals each day. That is why you should feed your horses two to three times each day.

● Horses have an extremely strong biological clock, especially when it comes to feeding. Feeding late or inconsistently can result in colic and other digestive upsets. Be sure to feed the same amounts at the same time every day.

● The horse's digestive system is adapted to a high amount of bulk and a low amount of concentrate. Be careful not to imbalance the ration too heavily toward grain.

● Feed each horse individually according to his specific needs. This avoids competition, fighting, and some horses gulping and getting too much while others get very little.

● Know exactly what you are feeding. Read and understand the feed tags of commercially prepared feed. Have your year's supply of hay tested for nutrient content if possible.

● Know exactly how much you are feeding. Feed by weight, not by volume. Feeding by volume contributes to overfeeding and wasted money. Feed hay at an approximate rate of 1.5 to 1.75 pounds per 100 pounds of body weight. This means that a 1,000 pound horse will require about 15 to 17.5 pounds of hay per day. It is best if you weigh hay at each feeding or you can weigh several flakes of the hay you are feeding to determine the average weight of a flake. Flakes (also called fleks, leaves, slabs, or slices) can vary from two to seven pounds depending on the type of hay, moisture content, how tightly the hay was baled, and the adjustment on the baler for flake thickness.

The approximate weight of your horse can be determined from the following table:

HEART GIRTH IN INCHES	WEIGHT IN POUNDS
54	492
56	531
58	596
60	664
62	722
64	785
66	852
68	902
70	985
72	1065
74	1220
76	1265

Feed grain to young, growing horses, horses in hard work, and lactating broodmares. Because grain should be fed by weight, not volume, never use a scoop to measure, since it is far too inaccurate. Oats are much lighter than corn, for example, so a quart of oats will weigh far less than a quart of corn. Here is weight of a quart of some of the grains commonly fed to horses: bran, ½ pound; oats, 1 pound; barley, 1½ pounds; corn 1¾ pounds.

The energy values of grains, expressed as total digestible nutrients (TDN), vary greatly. The TDN of oats is 65 percent; barley is 72 percent; and corn is 85 percent. This means that a pound of corn contains nearly a third more energy than a pound of oats. It should be determined how much additional energy the horse requires in addition to the energy provided by the hay he receives.

● Make all changes in feed gradually. Whether it is a change in the type of feed or in the amount being fed, make the changes in small increments, and hold the amount at the new level for several feedings. If you are feeding 4 pounds of grain per day and want the horse to have 6 pounds per day, increase to 4½ pounds and feed that for at least two days. Then increase to 5 pounds for two days, and so on. If you are making a change in hays, feed one part new hay and three parts previous hay, hold for two days, and then feed half and half for several days, and so on.

● Be aware that a pasture- or grain-fed horse that is brought suddenly into work can suffer azoturia, or "tying-up." This usually afflicts a horse that is vigorously exercised after a period of rest (several days or more), during which the feed has not been decreased. When the horse is forced to exercise, excess lactic acid accumulation in the muscles results in tenseness and soreness, often preventing the horse from moving at all. To prevent such a situation from happening, decrease your horse's grain ration when he is not being exercised for two days or more. When you do resume work, be sure the horse is given a very thorough warm-up and cool-down.

● When turning your horse out to pasture for the first time every spring, be sure he has had a full feed of hay. Limit grazing to one-half hour per day for the first two days, then one-half hour twice a day for two days, then one hour twice a day, and so on. Keep a close watch on horses that are on pasture as they can quickly become overweight or suffer the devastating condition laminitis (founder) from too much rich or green feed. If a horse has been off pasture for a week or more, reintroduce him to the green feed gradually.

● Be sure it is impossible for a horse to get to the feed in your storage areas. Horses do not know when to stop eating and can literally "eat themselves sick." An excess amount of grain can cause colic or laminitis.

● Horses have sensitive digestive systems. Do not feed a horse immediately after hard work, and do not work the animal until at least one hour after a full feed.

● Feed the highest quality hay you can find. Take the time to shop around and become familiar with the characteristics that constitute excellent hay. (See the chapter on pastures and hay fields for more information.)

● Be sure feeders are clean and safe. Do not let feed accumulate in bottom of feeders. Moldy or spoiled feed can create problems for your horse and large veterinary bills for you. Routinely check all feeders for sharp edges, broken parts, loose wires or nails, or any other hazard.

● Do not feed horses on the ground where they may ingest sand or decomposed granite along with their feed. This can cause sand colic, a dangerous type of impaction. Concrete pads or rubber mats in feed areas may be helpful in preventing sand colic.

● To take the edge off an overeager horse's appetite, consider feeding hay first and following it ten to twenty minutes later with grain. Horses that gulp or bolt their grain can suffer choke, colic, or poor feed utilization. To force a horse to eat its grain more slowly, mix large hay wafers, cubes, or "cakes" in with his grain ration or leave several softball-size smooth rocks in his grain feeder. A large, shallow grain feeder will cause a horse to eat more slowly than will a narrow, deep grain feeder such as a bucket.

● Balance your horse's ration by providing, free choice, trace mineralized salt. This contains sodium, chloride and, usually, iodine, zinc, iron, manganese, copper, and cobalt.

● Depending on the horse's age and type of feed, determine if calcium and phosphorus need to be supplemented and in what ratio. If calcium is deficient, limestone can be added to the grain. If phosphorus is low, monosodium phosphate can be used as a source. If both calcium and phosphorus are low, di-calcium phosphate can be used.

● If you plan to be away from home a lot and are considering using self-feeders, be aware that there can be a higher incidence of colic and laminitis with self-feeders than if feed-

A safe, custom-made hay and grain feeder in an outside pen. The feeder is flush with the inside of the pen, and feed is loaded in from a top lid. Feed pulled out by the horse falls on a rubber mat, that is anchored to the ground by two metal rods.

Hay can also be fed in hay racks. The tray will catch the leaf that breaks off.

ing is more closely and regularly monitored. Self-feeders are designed to feed processed feeds and grains, not long hay. Parts of the ration may separate, delivering inconsistent ratios of grain and ground hay. Complete feed pellets may crumble and become dusty. Refused feed may become wet and moldy, increasing the chance of digestive problems. In addition, self-feeding systems may encourage some horses to overeat and become fat.

Water. Horses drink between 4 and 20 gallons of water a day. Their water should always be available, clean, and of good quality. (See chapter 13 on water for more information on quality.) A horse's water intake will increase with environmental heat, exertion, lactation, increased hay ingestion, some illnesses, and increased salt intake. Horses drink less water in extremely cold weather and during some illnesses.

In very cold weather, the horse's body temperature will not drop as much from drinking "warm" water as it would if forced to eat snow or drink from a frigid pond. However, many horses will not drink artificially warmed or too-hot water. If you are in a cold climate and do not have heated watering devices, the best bet is to draw fresh buckets from the tap or hydrant several times a day to offer each horse. Horses prefer water at 35 to 40 degrees Fahrenheit.

It can take quite some time for a horse to get its fill. Horses drink about one-third of a pint per swallow or one gallon in about a half minute, coming up for air after about ten swallows or so. A horse drinks by closing his lips and creating suction with his tongue. A bridled horse may take much longer to get the water he needs as the bit prevents him from sealing off the corners of his mouth. A snaffle bit does not pose as much of a problem in this regard as does a curb.

A horse should be watered frequently during a long workout. A hot horse that has had water withheld should not be watered fully after hard exercise, however, as it could result in cramping, colic, or water founder. The horse should instead be hand-walked and offered small sips until its pulse, respiration, and temperature have normalized.

Because horses have such a delicate sense of smell and taste, they can refuse a new water (especially away from home), even though it may be perfectly safe to drink. To prevent this from happening, you may need to flavor the current "home" water for a while before changing to the new water, and then flavor the new water with the same agent. Substances that are useful in this regard are oil of peppermint, oil of wintergreen, or molasses. Use very sparingly, just a drop or two per bucket.

If a horse has access to clean, pure water at all times, it is unlikely that he will suffer dehydration. To assess your horse's level of hydration, use the pinch test. Grasp a fold of skin on the horse's neck and pull it away from the muscle. Let it go and see how quickly the skin returns to its flat position. The more pliable and resilient the skin, the higher the level of water in the horse. The skin should return within two seconds to its original position. If the return is slow, there is moderate dehydration. If a "standing tent" of skin remains, the horse is dehydrated and a veterinarian should be consulted.

Shelter and protection. It is not necessary to have an air-tight, heated barn for horses; in fact, that is one of the unhealthiest environments in which a horse can live. (See chapter 7 on barns.) A horse's shelter requirements are pretty basic—a place to get out of the wind and hot sun and a way to avoid getting or staying wet during cold weather. Of course, the more complicated we get with winter hair coat clips, blankets, and so on, the more complicated the shelter/protection formula becomes. Refer to chapter 4 on management styles to see what level of shelter will work best for you. Then read chapter 16 on daily routines to see the amount of labor that is involved in keeping a horse in a stall.

Exercise. Exercise is essential for health and for the proper development of young horses. It maintains a balance between feed ingested and bodily waste and is essential for body building and repair. Exercise, in contrast to the progressive training effects of *conditioning*, is often referred to as *maintenance*. The term "idle," used when formulating rations, does not indicate that the horse is not allowed or encouraged to exercise but that the horse is not being used for regular, strenuous work at that particular time. All horses of all ages need exercise every day— either a daily ride or a minimum turn-out of two hours in a large pen or pasture.

A regular exercise program invigorates the appetite, tones muscles, increases lung and heart capacity, and helps develop reflexes and coordination. Exercise increases circulation which increases the activity of the skin and lungs, which in turn helps remove body heat and the waste products (lactic acid) of exercise. Exercise aids in the development and repair of tissue and improves the quality and strength of bones, tendons, ligaments, and hoofs. Regular stress creates dense, stress-resistant bone. Exercise also conditions and stretches muscles and tendons, resulting in less chance of injury and lameness. Allowing horses to play in moderately soft footing can help develop elasticity in tendons.

Horses that are allowed ample exercise rarely develop vices such as pawing, stall kicking or wood chewing, which are often results of boredom.

Adult horses take the largest portion of their exercise at the walk but young horses, testing their physical limits, have explosive outbursts at all gaits. Since foals and yearlings are characteristically insecure, vulnerable, excitable, and unpredictable, it is essential to

Rocks and trees can offer adequate shelter provided the horse is in good health and well fed.

provide them with a safe place to exercise. And since a horse's vision is less than perfect, it is important that any exercise area is safely fenced and free from hazardous objects. Footing should be soft, but not excessively deep. Hyperextension of the fetlock in deep sand can do permanent damage to tendons.

For the days on which you cannot ride, free exercise is the least labor intensive and most natural way of providing exercise for your horses. There are other options, however. *Ponying* (leading one horse while riding another) is a good choice, especially for young horses. Ponying can start in an arena but can be expanded to include work in open spaces on varied terrain. Ponying a young horse on the surface that he will be worked on when he is an adult provides an opportunity for specialized adaptation of tissues. And the variety in scenery and experiences during ponying is good for any horse.

Longeing—working a horse around you in a circle on a thirty-foot line—is an option for horses over two years of age. Due to the uneven loading of the legs associated with repetitious work in a circle, a younger horse may suffer strain.

Electric horse walkers are useful for occasional sessions but should not be viewed as the mainstay of the exercise program. Thirty minutes of walking once or twice a week might be a good alternative on busy days. Depending entirely on a walker for exercise, however, encourages a stiff carriage, resistance, laziness, and boredom.

Treadmills can also be used for an occasional workout, providing that the horse is gradually conditioned to the work and carefully monitored for signs of stress. A continuous climb at the 5-to-7 degree slope characteristic of most treadmills is extremely fatiguing. A workout using a treadmill is accomplished in about half the time required for most other forms of exercise. If a young horse is asked to perform on the treadmill for even a few minutes beyond his physical capabilities he might become injured or sour

towards work. Treadmills are used successfully for muscle development, particularly of the forearm, chest, stifle, and gaskin.

Horses are naturally good swimmers. When paddling through water, they receive a good deal of exercise without traumatizing the joints.

Rest. Horses rest in one of three positions: standing, sternal recumbent, and lateral recumbent. A unique stay apparatus and a system of check ligaments allow a horse to sleep while standing up. And this is how most horses do, in fact, rest. In order for a horse to have quality rest while dozing on its feet, it simply needs a comfortable place to stand—one that is relatively level and free from weather extremes, noise, light, insects, and anything threatening.

On a sunny day or after a particularly hard work-out, your horse may lie down to rest. First, perhaps, he will try the sternal recumbent position by simply kneeling, then tucking his hind legs under his body and lying on his belly. Horses often take a snooze while tucked in such a cozy little ball. If suddenly startled, however, most horses can rise from this position in an instant because the hind legs are under the body ready to push it up.

If a horse is very relaxed and unthreatened, it may roll over from the sternal recumbent position and lie flat on one side, extend all four legs, and lay its head and neck on the ground. It takes more time to get into a "red-alert" position from this lateral recumbent position than from the other two, but the horse can be up more quickly than you might imagine.

Most horses cannot resist rolling, which is a form of self-grooming, when in one of the latter two sleep positions. When any horse lies down, whether for rolling or sleeping, it first bends fetlocks and knees and falls to the ground. Note that if the ground is hard or abrasive, the horse may suffer chronic abrasions on the front surfaces of these joints.

Sometimes a companion animal will help alleviate a single horse's loneliness. COURTESY OF DALE FORBES BORMANN

It is perfectly natural for a healthy, fit, sound horse to lie down occasionally, but if your horse spends a large amount of its time lying down, it may have hoof or leg problems and you should get professional advice and help.

Companionship. A social creature, the horse would be very lonely without interaction with others of its species. Horses do not need actual physical contact with one another, but must be near enough to see, smell, and hear each other. The more you interact with your horse, the more you will provide him with some of the aspects of companionship that he would normally get from other horses. This point must not be taken too far however, as you are not a horse and he is not a human. If you keep things in perspective, you can develop a healthy partnership with your horse and you will both be better off for it.

Veterinary and farrier care. If you follow good management practices, your veterinary and farrier bills will be minimal. Find a good veterinarian and a competent farrier, and then don't alienate them with inconsistent scheduling. Earn their respect by being the very best horseman you can and they will be there to help you when you need them the most. Schedule routine visits by your farrier every eight weeks or less. Unless you do some of the work yourself, your veterinarian will need to attend to your horses every two months for deworming, vaccinations, dental care, and other routine procedures. Don't try to cut corners on hoof or health care. Your horse's well-being is at stake and you can't afford to risk it.

A little something extra. I have talked about the needs and requirements of the horse but sometimes you want to do just a little something extra to show your appreciation to your special animal.

At the top of any horse's list is more food! One common extra treat added to the horse's diet is a molasses-based protein block. Although horses dearly love these large "candy bars," I want to caution you about allowing your horses unlimited access to them. Sold across the country under hundreds of local feed mill labels, these blocks must be regarded as a supplement to the horse's normal diet. Under most feeding circumstances, they are unnecessary, so are an added expense and can be toxic if a horse gobbles them down, especially those containing mineral and vitamin supplements.

Comprised of grain products, molasses and possibly minerals and/or vitamins, the 40-to-50-pound cubes have a wonderful smell and a texture that entices horses to both lick and gnaw on them. Similar products are made for sheep and cattle, but contain a synthetic source of protein called urea. For horses, it is important to purchase the "premium" version, which contains protein from natural (plant) sources, such as soybean meal.

If your horse seems determined to finish a block in one session, you will have to play referee. Maybe you can roll the block into his stall or pen for a half hour each day. Be sure there is adequate water available, as even the small percentage of salt in most of the formulas will increase your horse's thirst. And if your horse is already receiving adequate protein in his diet, you should use something else to show your appreciation.

A safe and healthy palate-pleasing treat for your horse is an occasional carrot or apple. Rather than feed them from your hand, however, which may encourage future nibbling, just pop one in his feed tub for him to find later as a surprise. If your horse develops a serious addiction to carrots, vegetable farms will sometimes sell an entire pickup load for under $100. If you have a cool shady place to store them, they will most likely keep until the last one is fed. Carrots provide a welcome diversion to the

horse's normal ration and can be a healthy reward for good behavior. They are an excellent source of carotene, the precursor to vitamin A—usually the only vitamin that ever needs to be supplemented in a horse's diet. If your horse is not receiving green sun-cured hay, he may not be getting adequate carotene.

When the temperature dips severely, oatmeal makes an inviting breakfast in the house. Out at the barn on a chilly morning, your horses may also appreciate a warm grain mash. It takes a little practice and some testing to see what proportions appeal most to each horse. You are not limited to bran as the sole ingredient in a hot mash. Experiment with oats, cracked corn, barley, a handful of molasses, or a pinch of salt, and you are on your way to satisfying your horse's culinary taste (or at least enjoying the benevolent feeling you get from trying!).

Measure and mix the dry ingredients the night before and bring them to the house in a pail. When you plug in the coffee pot in the morning, heat a kettle of water for the mash. A 4:1 ratio of grains to boiling water is satisfactory for most horses. According to the equine palate, it is best to err on the dry side rather than the mushy side. Stir as you pour the water. Let the mash steep in a warm place for about thirty minutes. Check the temperature and serve. For a treat for yourself, take a hot mug of your favorite beverage out to the barn, find a warm spot to sit, and then listen to the contented slurpings of your appreciative creatures.

A social treat for your show horse or other stalled mount is a special time when he is allowed to be a plain ol' horse. Give him some time off in a large, interesting pasture. Plan it for a time when it's OK if he messes up his mane and tail and has a good roll. Many horses like nothing better than to nose around a pasture inspecting roots and sticks and tracing recent equine history. Contrary to human guidelines, horses see nothing wrong in being dirty or having their manes flop over to both sides of their necks.

For yet another treat, give your horse a vigorous body massage. It can be a one-time thirty-minute rubdown or a week of fifteen-minute body-stropping sessions. Body stropping is an isotonic muscle exercise. With a stable rubber or a wisp, pound the large muscle masses of the neck, shoulder, and hindquarter with moderate pressure to stimulate circulation and then follow with a sweeping motion. This encourages removal of both internal and external waste products. With your bare hands, massage your horse's legs in circular motions toward the heart. Give your horse a head rub and then finish with an ear massage. Pull the ears slightly out to the side and let your fingers slide off the tips. But beware, horses given such a body rub are likely to melt into a puddle at the cross-ties and might need to be carried off to their stalls!

If you are a one-horse family, you may wish to consider a stall companion for your horse. Some friendships just happen and do not have to be arranged. Cats, chickens, lambs, and dogs have been known to voluntarily take up quarters with a compatible horse. The daily treks and routines of both horse and companion provide interest and comfort for each other. Pygmy goats and other pets or small livestock can sometimes be successfully introduced to a lonely horse.

If your horse must spend a lot of time in his stall and, despite your good efforts, is sometomes bored, an innovative hanging stall toy might help him while away the hours. Stall toys can channel pent-up energies toward non-destructive play and help prevent vices. Commercial models are often huge plastic fruits or vegetables, but a cleverly painted gallon milk container works just as well. Experiment with hanging the toy from various heights. Be forewarned that if your horse becomes addicted to playing with a toy, you may see some physical changes in his neck...some desirable, some not so pretty. Another idea is to give a horse a sturdy beach ball to play with in a small paddock or indoor arena.

I think the best thing you can possibly do for your horse is to become the very best horseman you can possibly be. Take the initiative to learn all there is to know about horse behavior, training, riding, health care, and management. One way you can make an immediate hit with your horse is to make his work more pleasant, and a key way to do that is to become a more physically fit and athletic rider. Lose a few pounds if necessary, do those stretching exercises, strengthen your body, and become a working member of the team, not just a passenger. Schedule some lessons and take your instructor's advice. But then that is the subject for another entire volume!

First determine what type of supplement block a horse needs, then provide it free choice.

CHAPTER FOUR

Levels of Management

WHICH LEVEL OF MANAGEMENT BEST SUITS YOUR particular lifestyle and location? Do you have ample pasture but little time for daily horse care routines and plan to ride only on weekends? If so, then keeping your horses on pasture may be the best choice. If, on the other hand, you have limited space but have the time and desire to attend to stable chores and daily riding, then keeping your horse in a stall in the barn would probably suit your situation better. In between these two extremes are many variations of management schemes. For the sake of discussion, however, and to help you devise your own personal management plan, here's a look at the advantages and disadvantages of keeping a horse on pasture.

Keeping a Horse on Pasture

Advantages

Keeping your horse on a well-maintained pasture with one or more other horses is the most natural style of management. It allows free exercise, fresh air, sunshine, and socialization. Horses on pasture tend to stay healthier with fewer of the respiratory diseases that are seen with stalled horses. Pasture horses stay "legged up": that is, the tendons, ligaments, bones, and hooves usually receive more regular stress, becoming tough yet resilient, and are less prone to lameness from a misstep or slip. If a regularly ridden pasture horse is a fairly energetic self-exerciser, his respiratory and circulatory systems will retain a higher level of fitness than will those of the stalled horse who receives the same amount of riding but does not have the benefit of free exercise.

A well-kept pasture can offer excellent nutrition, especially minerals and vitamins A and E. Sunshine is a source of the precursor to vitamin D.

Pastured horses seem to have a better mental attitude than many stalled horses because they are allowed "to be horses." Pastured horses rarely develop vices. However, it should be noted that although turning a horse out on pasture may help cure a vice, there are many cases where a horse may continue the established undesirable behavior even while out on pasture.

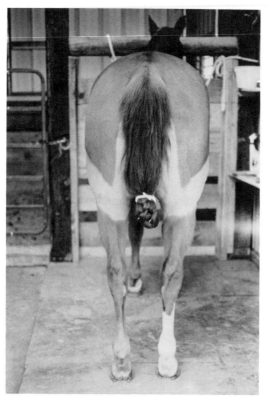

To prevent a long tail from balling up with frozen snow and ice, it can be braided up for the winter.

some months. This means you will either need enough pasture so that you can rotate through the various pastures all year, or you will have to use some confinement facilities for part of the year. If you leave too many horses on an inadequate pasture for too long it will be ruined by overgrazing. Overgrazing results in destruction of plant and root structure, takeover by weeds, and soil erosion.

Certain horse behaviors are also destructive to pastures. Pawing, especially with shod hooves, can damage plants. Horses tend to select different areas in which to graze, to defecate, and to congregate. This results in some areas being overgrazed, some being ignored because of fecal contamination, and others turning into bare dirt or mud holes.

Pastured horses are less handy to catch and ride. You must either teach them to come when you call or you must walk out to get them. If a horse has not been properly taught to be caught, he can play the "catch me" game which can be very time consuming and frustrating. (See chapter 16 for more information on handling pasture horses.) When horses are turned out to pasture and not handled very often they tend to revert to the "wild" behaviors they had before training. The freedom can make many pasture horses feisty and headstrong.

Horses turned out for the winter grow a long coat. When such horses are exercised, they tend to sweat profusely and lather and are very difficult to cool out properly. A wet coat is a poor insulator against the cold and invites respiratory illness. When the long coat is wet, snowy, or muddy, it requires extensive grooming before riding. And this is somewhat of a "Catch 22" situation because when you groom a horse who is spending the winter out on pasture, you remove sebaceous secretions, a waxy build-up on the skin, which is essential for protection from moisture and wind.

Horses kept in groups rather than individually tend to run more and fight (even though sometimes in play) with each other.

Managing horses on pasture can be less expensive than keeping horses in stalls. This depends on the availability and cost of land versus the cost of facilities. Keeping horses on pasture is far less labor intensive than keeping them in stalls. This is not to say that pastured horses do not require a daily check (see chapter 16 on routines), but watering, feeding, and manure management will often be greatly simplified.

It probably sounds as if pasturing horses is the way to go! In some instances, it certainly can be. But take a look at the disadvantages before you decide, and you will begin to see which management style is best suited for you.

Disadvantages

First of all, land is generally very expensive and horses can be very hard on it. They eat constantly. That means the pasture growth must be able to support the number of horses you intend to keep. But no matter how good a pasture is, it will need to rest for

Horses turned out for the winter grow very long winter coats.

Both situations lead to more injuries for horses on pasture. To minimize accidents, pastures must be well fenced and the fences must be maintained. This is both costly and time intensive.

Although pasture grasses offer good nutrition, horses on full pasture feed often tend to overeat, develop a grass belly, get out of condition, and become lazy. Grass founder, a severe inflammation of the hooves caused by a digestive overload, is an all-too-common occurrence with pastured horses. In addition, pastured horses eat from the ground, so they run the risk of sand colic and daily reinfestation with parasites. Sometimes sand (and other types of soil) will accumulate in the horse's gut and cause obstruction of the bowels.

You may find that keeping your horse on pasture works fine for years or you may soon feel constrained by the limitations that type of management imposes on your horse activities. Your pasture may need a rest or you may wish to get rid of your horse's round belly. In both situations you will need some sort of confinement for your horse—a stall or pen.

Keeping a Horse in a Stall

When it is essential for your schedule that your horse be clean and ready when you have time to ride, you will start dreaming about the benefits of stalls. When you want to delay and minimize the growth of the winter coat or speed up shedding in the spring, you will need to design facilities that will allow your horses to wear blankets.

As you become more involved in the routine chores of horse management, you will begin looking for a better area for your farrier and veterinarian to work. You will want a better space for grooming, clipping, bathing, and tacking up, too. And you may get tired of carrying your saddle from the house every day and start wishing for a tack room. And finally, if you have specific interests in a particular aspect of training or breeding, you will soon realize that you need specialized facilities for those endeavors.

Remember, as you move your horse from a pasture situation to a pen or a stall, you are changing his environment from one where he can make some choices that affect

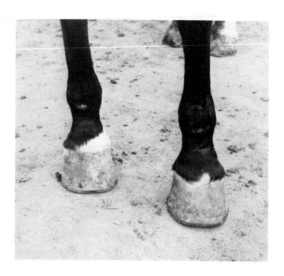 

Horses kept in pens without any bedding can develop sores on their knees, hocks, fetlocks, and elbows from the abrasion when they roll or lie down to rest. This horse required protective boots for the chronic sores to heal.

his well-being to a situation where he is totally dependent on you, his human manager. If you choose to confine your horse, you must realize that since he no longer has the room to move to keep warm, you must provide shelter, clothing, and exercise to compensate. Because grooming, clipping, and bathing remove the protective coating close to his skin, you must give him clothing and shelter to protect him from wind and moisture as well as increasing his feed if necessary. Since a stalled horse no longer has the freedom to choose his footing or the air he breathes, you must keep stall flooring and bedding in the best of shape and ensure that the air in the stable is fit to breathe.

SECTION TWO

DESIGNING YOUR ACREAGE

....

BOB AND SUE OUTGREW THEIR THREE ACRES at a time when mid-priced acreages were moving rather slowly. Their place stayed on the market for a year and a half. During that time, Sue looked at every single larger acreage available but found only one that might work. It was an old farmhouse on twenty acres. The pastures were improved and irrigated. The location was quiet, but there was easy access to a main highway. The price was almost too reasonable. Granted, the house needed some remodeling, but it could be made into a home with rich character. Since the piece wasn't the perfect answer to their dreams, Sue had not taken the time to check all details closely. When the buyers for their three-acre piece finally materialized, however, a quick closing was scheduled. Sue and Bob were pressured by the realtor to put down some earnest money on the twenty-acre piece, so they'd have somewhere to move to.

Ending A: They bought the farmhouse on twenty acres. During the first week, they noticed the water tasted unusual and the laundry was not getting really clean, prompting them to have the water tested. The results showed that the well water was so contaminated with nitrates and salts that not only was it unsuitable for human consumption, but it would be a severe hazard for their pregnant and lactating mares and for any horses under a year of age. They inquired about a water treatment system to remove the undesirable elements and found that the installation would cost them over $10,000 (more than the price of a new well) and the annual maintenance costs for filters and additives would be over $500 per year.

Ending B: They decided, however, that since they hadn't really found the ideal acreage, they would be better off leasing a place for a while and continuing their search. Sue picked up the Sunday paper and looked in the "Acreages for Lease or Rent" column. She spotted a listing for a fully fenced forty-acre piece with a house and a barn. When she and Bob went to see it, the owners told them it was also for sale. It had everything they were looking for and more. They made an offer on it with the contingency that the quality and volume of the water in the well was satisfactory. When the tests were completed they purchased their new place and lived there happily ever after.

CHAPTER FIVE

Choosing An Acreage

WHEN CHOOSING COUNTRY LAND, THERE ARE MANY factors to consider. Arrange the following items in an order that suits your location, financial situation, and goals. Be sure to seek competent, experienced advice on matters with which you are unfamiliar. Contact your county extension agents to check about soil, water, climate, and plant growth in the area. Visit other horse farms to get an idea of the advantages and disadvantages of various facilities in your locale. Specifically, consider the following:

Location

Before you even begin looking at properties, decide how far away from a town you and your family can realistically exist. How many miles and minutes are you willing to drive to your job(s) each day? Where is the closest school? Will you spend most of your time in the car ferrying your children into town for activities? Will you have convenient access to the services of a farrier, veterinarian, and feed store? Some of the most beautiful horse country and the best land bargains are a stiff commute from employment centers. Don't make the mistake

of falling in love with a place that would be perfect for horses but tough for the realities of human existence.

Determine a reasonable distance you are willing to commute daily and, using a compass, draw an appropriately sized circle on a map to identify the area in which to begin looking. Don't automatically eliminate a property outside of the circle, however, because its other advantages may result in a quality of living far beyond what is available inside the radius. If you are considering a remote location, make test drives to the property and see how much time traveling usurps. There is little sense in living in a wonderful spot if you can only occasionally see it in the daylight!

Climate

Although climate is somewhat predictable, you should be cognizant of small pockets of unusual weather. Such micro-climates can result in one property being ideal and one a few miles away being windblown, flooded, gloomy, or dry. Consider temperature, precipitation, wind, humidity, and length of growing and riding seasons. Being really familiar with an area before you buy is the only way to find out these things. I have seen one property sell almost every winter then come up for sale each spring when the majority of the land, including the homesite, stands under water for several months.

Water

The availability and quality of water are of utmost importance to you and your family as well as to the health of your horses (see also chapter 13). If the property does not have a well, determine if there are legal restrictions that prevent drilling one. If the property will require irrigation, find out if there are water rights included or available for purchase which will meet the needs of the land. Find out the depth of the groundwater table. This will determine if you will be able to do any underground construction and may affect the installation of a septic system.

Topography

Note which way the slopes face and what kind of natural or manmade structures may block the sun. In temperate climates, north-facing slopes or areas obstructed from light may mean increased heating costs, icy driveways and barnyards, increased wind chill and snow drifting, and a "later morning" in the winter. On the other hand, the same features may be advantageous in reducing heat on a warm-climate acreage. Moderately rolling grassy hills are desirable as they usually drain well and provide a good environment for the exercise and development of horses. A slope of two to six percent (two to six feet of rise or fall per one hundred feet) is ideal. A much greater slope than this might result in erosion problems and would probably require extensive excavation costs for any improvements you might add to the land. A slope much lower than two percent would offer a poor chance for drainage and result in wet, marshy,

boggy areas. Such areas, which are mosquito breeding grounds, contribute to the possible spread of Equine Infectious Anemia and other diseases. In addition, wetlands encourage the growth of disease-causing organisms and are very damaging to the hooves. See chapter 14 on sanitation for the effects of excess moisture.

Soil

Along with the slope, the type of topsoil will greatly affect surface drainage. A sandy loam is ideal. The soil should not pack or become excessively muddy for long periods; yet, neither should it be extremely sandy, or nutrients will be leached out of the soil by rain and melting snow. In addition, sandy soil results in more cases of sand colic. Clay and adobe soils pack hard, cup and retain the uneven surface, and are slippery when wet. Gravely soils lack nutrients and result in a greater number of hoof and leg injuries from the abrasive surface.

The topsoil and subsurface soil can be tested and evaluated. The soil profile will help you plan your planting, fertilization, and irrigation needs as well as deciding where to locate buildings and facilities. See chapter 12 on pasture management for more information on soil testing.

Other Natural Amenities

Since it takes so very long to establish a good stand of trees and since it is impossible to transplant mountains, streams, bluffs, or ponds, each of these natural extras should be considered when tallying a property's assets. Besides adding beauty, these features can satisfy a horse's needs for shelter and water. An added bonus to any land is its close proximity to public lands for riding.

Utilities and Services

Undeveloped tracts of land can be real bargains but, to determine this, check on the availability of utilities and the cost to get them to your building site. These include electricity, telephone, municipal water or a well, natural gas lines, and irrigation. If you wish to heat with propane, coal, or wood, check on their availability and on any regulations which may prohibit their use. Find out if the property is connected to a municipal sewage system and what the monthly charges are. If the property has its own septic system instead, be sure the sub-soil is permeable so percolation will take place, that the drainage field is sufficiently large, and that the tank is located where it can be cleaned easily. Inquire as to the location of the nearest landfill and of the availability of trash disposal services. Find out who maintains the roads and removes snow.

Find out who provides police service, fire protection, and ambulance service to the area. Will there be adequate water on the property for fire protection needs? Some of the answers may affect the fire insurance rates on your homeowner's policy.

Zoning and Other Legalities

Various city, county, and state regulations may affect what you can or cannot do on your future property. Restrictions may include but are not limited to where you locate your buildings, how many horses you are allowed to have, sanitation requirements, and the management of stallions. In advance, research the zoning classification and regulations that affect the land you are considering. The classification determines the permitted uses of the land. Places to look for information are the local planning department and the health department.

There may be further restrictions on your use of the land if the tract is a part of a homeowner's association in which membership is mandatory. While a homeowner's association may be limiting, if it is made up predominantly of horsemen it can offer protection for your horse interests as well.

Sometimes a group of landowners may initially unite for something simple such as the maintenance of roads that are not cared for with public funds. Some associations expand to adopt covenants that specifically define building styles and land uses of the members. In any event, it is a good idea to get a feel for your neighbors and the people who live in the area you are considering.

Buildings added to any property must usually comply with county and/or state building codes that outline required building specifications. See chapter 6 for more detail.

Roads and Access

A horse operation of any size usually requires trailering and trucking. Be sure the property has all-weather access for the delivery of hay and grain, building materials, and gravel and cement. It will also be necessary to get a horse trailer in and out during all seasons in the event of emergency.

If applicable, find out who owns the access road to the property. If the road crosses someone else's property, determine if there is a permanent easement for it. If there is not, before you purchase, you should draw up a permanent legal right-of-way. Also, find out what easements are attached to your prospective property such as roads, trails, utility lines, and irrigation ditches. Determine what rights these easements give other parties. This information will be found on the property deed or in the county recorder's office. While you are checking, see if there are rights that go along with the property, such as mineral or water, and see if there are any recorded rights that others have on the property, such as hunting or fishing.

Pollution

Air and water will vary in quality. Even if a parcel is far from a large population center, it may have one or more of the following hazards: smoke from factories, sawmills, or power plants; dust from mining or gravel operations; odors from agricultural operations, landfills, or sewage treatment facilities; noise from industry, highways, or airports; radioactivity from ore sites or mine tailings; or toxic waste from chemical dumps.

Be sure to check the source of the local waters. Water may contain lead or aluminum or, in a highly farmed area, agricultural run-off, which could be high in nitrates.

Number of Acres

If you can afford it, buy more acres than you think you will need. First of all, buying a five-acre piece with the thinking that you have a five-acre pasture is false. An acre or two is easily taken up by the house and yard and the barn and the barnyard. Secondly, although you may start out with one or two horses, you soon may find yourself expanding the herd. You may wish to raise a foal or buy an extra horse or two so that everyone in the family can ride together or so there is an extra horse for visitors. When planning your needs, ask the local agricultural agent what the carrying capacity for pasture land is in the area. You may plan for an acre per horse with improved, well-irrigated pasture but may need up to twenty or more acres per horse of dry rangeland. Finally, check to see if the property has ever been surveyed and if the boundaries can be easily located.

Pests

Some locations aren't very inviting for either humans or horses. Spots that are heavily populated with flies, ticks, other bugs, poisonous snakes, or scorpions will provide constant management problems and in some cases disaster. Don't let information such as this scare you, but check into it ahead of time so you are not disappointed.

The Future

Look around the prospective property for several miles and try to determine where future construction or other changes may take place that could affect the quality of water, air, or the aesthetics of the place. Your local (county) planning commission can tell you of future plans and variances that have been permitted and may affect your decision. Find out what the laws are regarding subdividing property or adding rental structures such as mobile homes.

While you are thinking ahead, consider the resale history of the property. A parcel that has been on the market for quite some time before you buy it will likely take a similar amount of time if you need to sell it. It may be a good value but so unique that it suits only a very few potential buyers.

Price

Price is often the first priority on a property shopping list, out of necessity, but don't forget to consider the factors that affect the price so that you are evaluating it in a proper perspective. First of all, compare the price of similar land that has recently sold. This is public information and can be found in the county recorder's office. When comparing two parcels, one may seem like a better deal when in fact its final cost will be much higher once equivalent improvements have been made. Calculate the cost of financing, utilities, and essential improvements. Find out the yearly taxes and if the property is located in any special taxing/service districts. If you are planning to pursue your horse involvement as a profit-making business, find out if the land is eligible for agricultural status and what you must do to qualify. And be sure to have enough cash remaining after the purchase for the horse facilities!

CHAPTER SIX

Designing the Layout

··

NOW COMES A VERY ENJOYABLE ASPECT OF YOUR endeavor. Here's where you can test all of your dreams and ideas on paper, where you can build an ideal horse farm, or where you can improvise and innovate with existing facilities. Country folks make an observation regarding the activities of newcomers to the area, as they note the progress of property improvements. If the animal facilities show a marked priority over the home, the person is a true farmer or rancher at heart!

Take plenty of time at the planning stages because you are going to have to live with your decisions. It's far cheaper to make mistakes on graph paper than with building materials! The larger the scope of your endeavor, the more I would urge you to consider hiring a professional planner, one who is very experienced with horse facilities.

First you must identify your goals. Keep in mind that your needs are very likely to expand. Try to determine exactly how many horses you wish to keep with what style of management, and what kind of training facilities you desire. Do some comparative shopping for building materials or for bids on the work. Finally, confer with your budget and make compromises if necessary.

No matter if you are starting with a bare tract of land or with a functional farmette, make your future plans with natural principles and conservation practices in mind. Look at the lay of the land, the soil, the weather patterns, the wildlife, the plant life. Consider the natural forces as you make your plans. Don't fight the runoff from a slope by locating your pens at the base of it. Don't subject your roofing to the effects of a wind tunnel by poor building placement. Enter landownership with the idea of improving it, not squeezing all you can out of it. Pasture improvement will be discussed in a later chapter but there are many other ways in which you can improve your land. Getting involved in water conservation, soil protection and erosion control, brush and tree management practices, and plant and wildlife protection will improve your land and add to your quality of living. You may even be eligible for cost-share assistance on a variety of programs administered by local, state, and federal agencies.

If you are starting with a piece of bare land (or one with just a house) you have a big job ahead of you, but you can design things exactly as you wish. Plan your facilities in

- Barn(s) with stalls

- Runs, pens, paddocks, pastures

- Storage for feed, bedding, machinery, tack, and other equipment

- Training areas: round pen, arena, track, walker, treadmill, pool

- Work areas: grooming area, wash rack, shoeing and veterinary area, breeding shed, laboratory, office, tack room

- Driveways, walkways, parking areas

- Shelter belts, wind breaks, wildlife areas

- Water and other utilities

relation to the improvements already there, such as residence, utilities, and fences. If there is no source of water on the land, the very first step is to drill a well in a location that will be convenient to the residence and the horse facilities. Take the natural features — the trees, the rocks, streams, and hills — into account, so that you don't lose what they have to offer but can incorporate it into your scheme.

If you have an operating small farm or a partially developed acreage, you must evaluate the existing facilities. Are they suitable as they are or do they need modifications? Are they usable temporarily as transition facilities? Do they have salvageable parts but require extensive renovation? Will the renovation cost more than a new building would? Would it be impractical to remodel the existing facilities, and should they be torn down instead?

After you read this book, study as many existing horse facilities as you can, especially those in your locale, and take notes and/or photos as to what you like and don't like. Make some initial sketches of what you want and show them to knowledgeable people in agricultural extension and in the building and construction business, as well as to other horsemen.

Facility Goals

When putting all of the pieces together, keep the following goals in mind.

Safety. Facilities should be strong and well designed with horse behavior principles in mind.

Convenience. Everyday activities should flow efficiently. Buildings should be placed so as to save labor and time. Locate water within easy access to the places it is needed.

Protection. Keep horse comfort in mind and provide protection from sun, wind, wet, cold, and insects.

Storage. Always plan for more storage space than you think you will need.

Economy. Without sacrificing quality, look into alternative materials and plans. Don't scrimp, however, on the finished dimensions of buildings or access lanes. The layout often takes more space in reality than it looks like it will on paper.

Flexibility. Keep some degree of adaptability in mind as you plan. Always leave room for expansion on to your buildings. You may shift the emphasis of your acreage or may need to sell it to someone without horses.

Locating Major Buildings

When locating your major buildings, consider the following. Plan for maximum sun in the winter and maximum shade and breeze in the summer. Check with the local weather authority to find out what the prevailing winds are during the various seasons. Go to the site itself during each season, especially winter, to determine which way the buildings should face. In the United States, prevailing winds usually come from the north, west, or northwest, so most farm buildings face the south or southeast.

Other buildings, trees, rocks, and slopes can have an effect on your proposed building site. They can obstruct the light, change the flow of air, causing draft or vacuums, and contribute excess runoff to the new building site.

Locate your buildings on dry ground, preferably high ground with a berm built up around the walls if necessary. To lessen excavation or fill dirt costs, find as flat an area as possible. Ideally there should be a 2-to-6 percent slope away from the building in all directions for surface drainage. The building floor should be eight to twelve inches above the outside ground level. If the building is located on a hill, a diversion ditch can be dug around the back side.

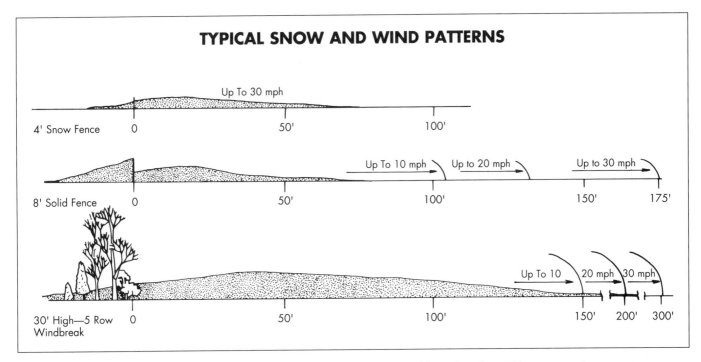

TYPICAL SNOW AND WIND PATTERNS

Up To 30 mph

4' Snow Fence   0   50'   100'

Up To 10 mph   Up to 20 mph   Up to 30 mph

8' Solid Fence   0   50'   100'   150'   175'

Up To 10   20 mph   30 mph

30' High—5 Row   0   50'   100'   150'   200'   300'
Windbreak

Ensure that there will be good subsurface drainage, especially for stall areas and runs, by having the subsoil evaluated. For a building location, a sandy or gravelly subsoil is preferred over clay or adobe soils. If necessary, have the site excavated. Refill the hole with large rocks, small rocks, road base or limestone and then let the site settle for several months before beginning construction.

Be sure that all key buildings have all-weather access for the delivery of building materials and eventually for hay, grain, bedding, etc. Plan for ample space to turn large trucks and/or trailers around. Assure that routine chores are possible without a great hardship during all seasons.

Locate key buildings close enough to the house for security and convenience yet far enough downwind so that flies and odor do not invade the residence. Formulate your fire plan as you plan your facilities. (See chapter 15 on fire safety.)

Make the appearance as nice as possible without sacrificing the functional aspects of your layout. Remember, plan for safety, efficiency, and convenience.

Choosing Your Builder

If you are not constructing your own facilities, either you will need to contact and contract various builders for the different aspects of your plans, or you will have to hire someone to do this for you. If you hire a representative, be sure that this person is very familiar with horse needs and facilities.

When looking for a professional tradesman, be sure to ask for references of previous clients. Take the time to look at the past work of the builder and talk to the people who hired him. Ask what type of warranty comes with a building or project. It is important to be very comfortable with the person(s) you hire.

So that you have an equitable means of comparison between bids, you or your representative should ask to see detailed building plans and/or specifications from every contractor you are considering. This will assist you in determining the quality of materials and work you can expect. You will need to devise a list of specifications you want in your barns, other buildings, fences, and any other facilities you will have built. When considering the construction of a pole barn, if you want an indication of joint strength, for example, you will need information on the system of bracing used and the location, number, and size of bolts or nails used to hold the braces together. Refer to the sample list of items to check on when getting an estimate for a simple pole barn. Similar checklists can be made for any

Diagram #1a
Windbreak Planning.

Wind Patterns: With a 40 m.p.h. wind from the left, velocities will be reduced to about those shown. For other speeds the reductions will be proportional.

PROTECTING OPEN-FRONT BUILDINGS

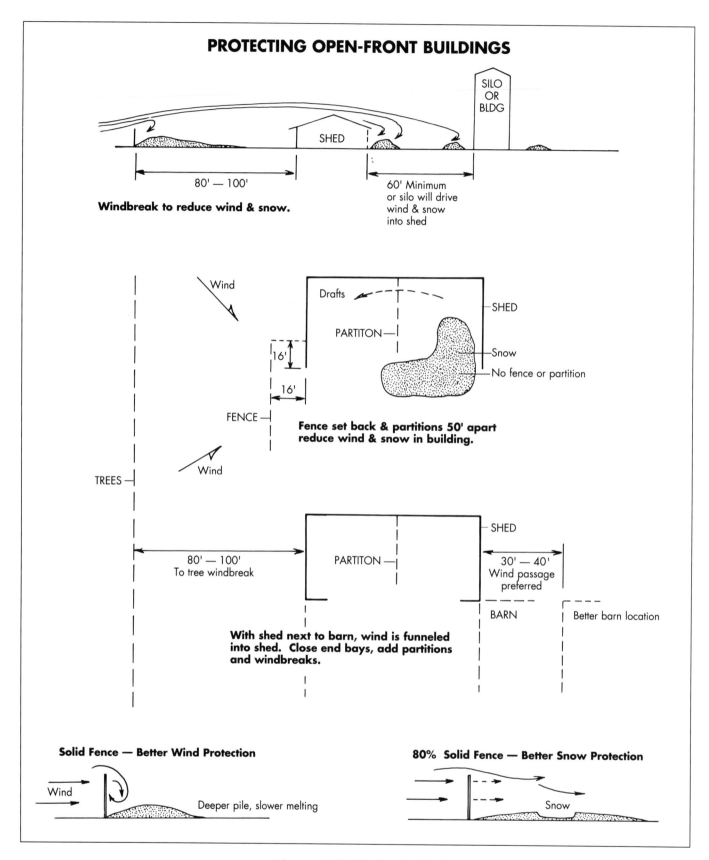

Windbreak to reduce wind & snow.

80' — 100'

60' Minimum or silo will drive wind & snow into shed

SILO OR BLDG

SHED

Wind

Drafts

PARTITON

SHED

Snow

No fence or partition

16'

16'

FENCE

Fence set back & partitions 50' apart reduce wind & snow in building.

Wind

TREES

80' — 100' To tree windbreak

PARTITON

SHED

30' — 40' Wind passage preferred

BARN

Better barn location

With shed next to barn, wind is funneled into shed. Close end bays, add partitions and windbreaks.

Solid Fence — Better Wind Protection

Wind

Deeper pile, slower melting

80% Solid Fence — Better Snow Protection

Snow

***Diagram #1b* Windbreak Planning.**

Orientation: Local experience is the best indicator on the distance facilities should be placed from shelterbelts. Generally, shelterbelts should be 100'-300' away from protected areas. The shorter distance is suitable where snow accumulation is less severe.

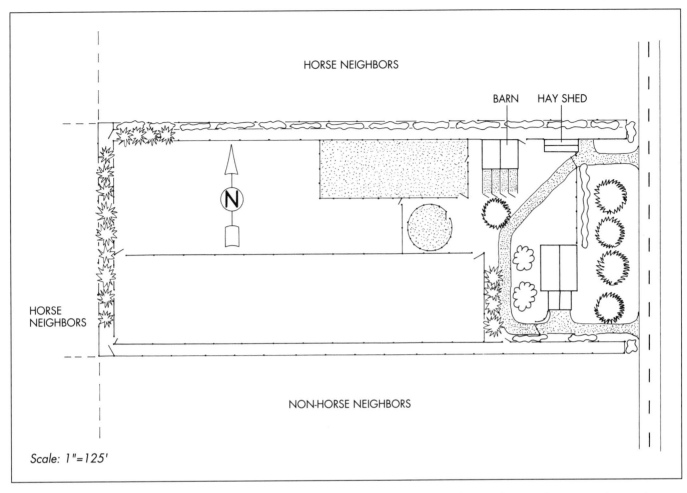

HORSE NEIGHBORS

BARN   HAY SHED

N

HORSE
NEIGHBORS

NON-HORSE NEIGHBORS

Scale: 1"=125'

Sand in pens
Gravel on
driveways

Pine Tree
or Fir Tree

Elm Tree

Fruit Tree

Cottonwood
or Willow
Tree

Maple Tree

Wire Fence

Board Fence

**Key for layouts
(Drawings 1-7)**

Drawing #1

RESIDENTIAL SUBDIVISION: 5 Acres,
300' (road frontage) x 726', developed from bare land.

House: Single story, 30' x 60', 1800 square feet, with two car (24' x 24') garage.

Barn: Four-stall, tack room, wash rack, 36' x 48', with four 12' x 32' runs. Runs are located off the gable end of the barn to avoid roof run-off. Runs are located on the south side of the barn for winter drying and warmth and so that they are visible from the house.

Hay shed: 16' x 48' which stores 600 standard size 70 pound bales. Hay shed is located with easy access for loading and unloading and separate from other buildings for fire protection.

Round pen: Post and plank, sand, 60' diameter. Visible from house for trainer safety.

Arena: Post and board, 70' x 200'. Electric fence can be added to pasture side of arena to prevent chewing.

South Pasture: 110' x 500'.

North Pasture: 140' x 460'.

Perimeter Runs: North and South are 16' x 680'; West is 20' x 300'. 16' gates and ends of run can be

opened to provide access for mower or horses. Long runs can be used as part of the rotational grazing and to separate horses, providing insulation between neighboring animals when they are present.

Landscaping:
Hedge along entire north border provides windbreak and visual privacy. Electric fence in north run protects hedge from horses.

Evergreen trees on west and north borders provide shelter for horses.

Gates: Two driveway gates ensure that loose horses cannot get onto roadway.

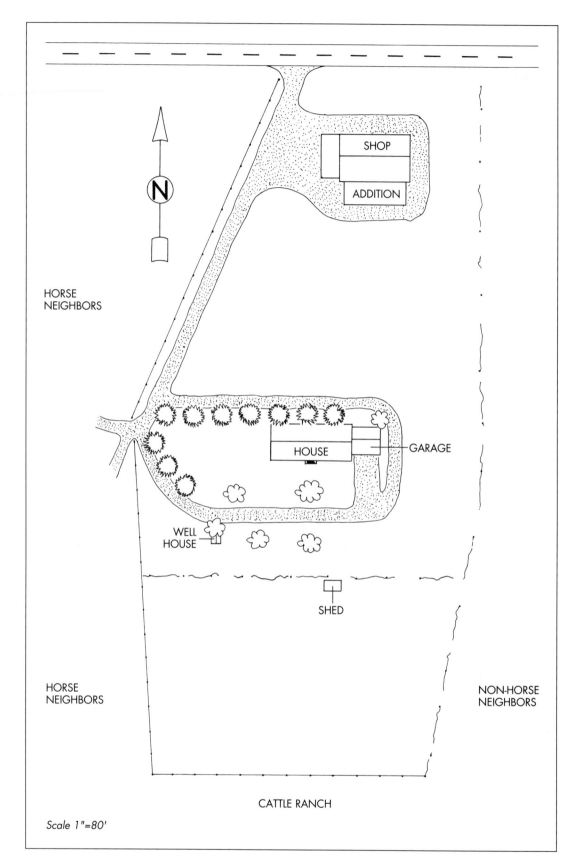

Drawing #2

BEFORE: 3.5 Acres, irregular shape, suburb of mid-size city.

House: 30' x 70' with attached two-car garage.

Shop: 60' x 80' with 20' x 75' addition.

Well house: 5' x 8' with heat tape on pipes.

Fencing: Marginal.

Horse facilities: None.

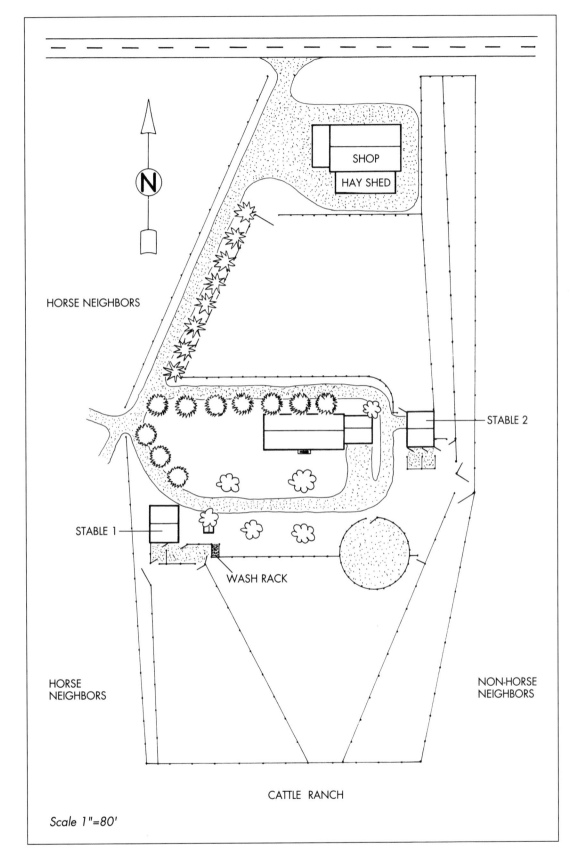

SHOP

HAY SHED

N

HORSE NEIGHBORS

STABLE 2

STABLE 1

WASH RACK

HORSE
NEIGHBORS

NON-HORSE
NEIGHBORS

CATTLE RANCH

Scale 1"=80'

Drawing #3

AFTER: 3.5 Acres.

Shop: Floor of 20' x 75' shop addition lined with pallets and used to store 1000 bales of hay.

Well house: Installed heater with thermostat to prevent freezing.

Shed: Removed.

Stables:
Stable 1: 24' x 28' (see Drawing #13 for details); two 12' x 16' stalls open to the south with pea gravel runs; 12' x 24' tack and feed room.

Stable 2: Same as Stable 1 except 12' x 24' tool and feed room instead of tack.

Wash rack: Built on existing concrete pad which was located in close proximity to water source (well house).

Pens:
66' diameter round pen, post and plank, sand footing. The central south pen separates the west and east pens.

Runs:
The long east runs can be joined to form one run the length of the property. A run insulates the north pasture from the east run to separate horses if necessary. The small run on the southwest border acts as insulation from neighbors' horses for health and safety reasons.

Fencing:
Cattle fences (barbed wire) were lined on inside with woven wire utilizing existing fence posts and were fortified on top with electric wire. All new fences are smooth wire or board.

Landscaping:
Pines along driveway for privacy, wind-protection, and shade.

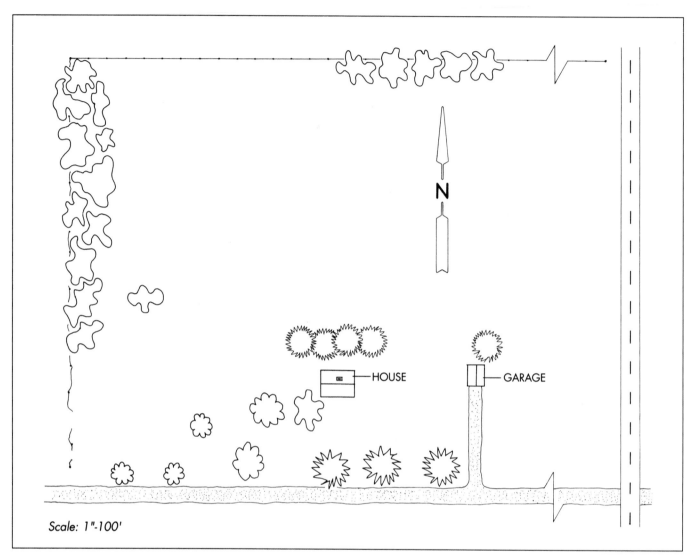

Scale: 1"-100'

Drawing #4

BEFORE: 10 Acres, farm community.

House: 30' x 40'.

Garage: 20' x 20'.

Layout: Two adjacent five-acre squares (467' x 467') with driveway to house approximately at center line.

Roads:

County gravel road frontage along south border.

County paved road frontage along east border.

Fencing: One good fence along north border.

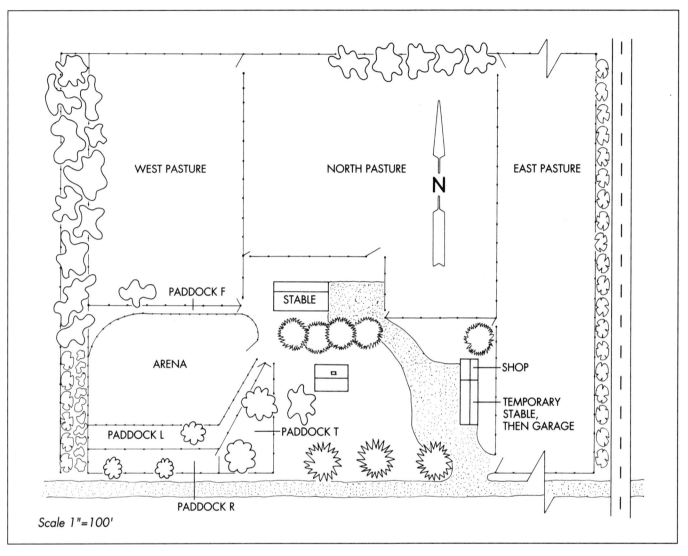

Scale 1"=100'

Drawing #5

AFTER: 10 Acres, farm community.

Garage: Expanded to 20' x 70' and used temporarily as stable and shop, then later as shop and garage.

Arena: 140' x 120'

Stable: 30' x 60'. See Drawings #10 and 11 for details.

Pastures:
West pasture: 170' x 290'

North pasture: 270' x 300'

East pasture: 467' x 467' (5 acres). Used as hay field for first cutting, then grazed the balance of the year.

Paddocks:
Paddock T (Triangle): 100' x 60' x 75'.

Paddock R (Rectangle): 25' x 140'.

Paddock L (L-Shaped): 25' x 225'.

Paddock F (Flask): 10' x 170'.

Landscaping:
West wildlife strip: 30' x 300'.

Maples planted along entire east border were all killed by county spraying. Post signs that say: NO SPRAYING!

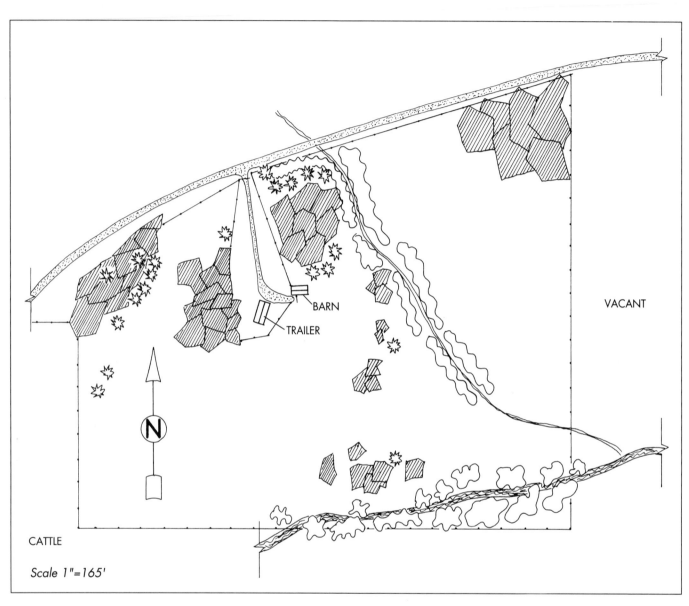

VACANT

BARN

TRAILER

CATTLE

Scale 1"=165'

Drawing #6

BEFORE: 20 Acres, ranch country.

House: 20' x 50' trailer.

Barn: 24' x 48' metal shell.

Fencing: Three strand smooth wire and metal post around entire perimeter.

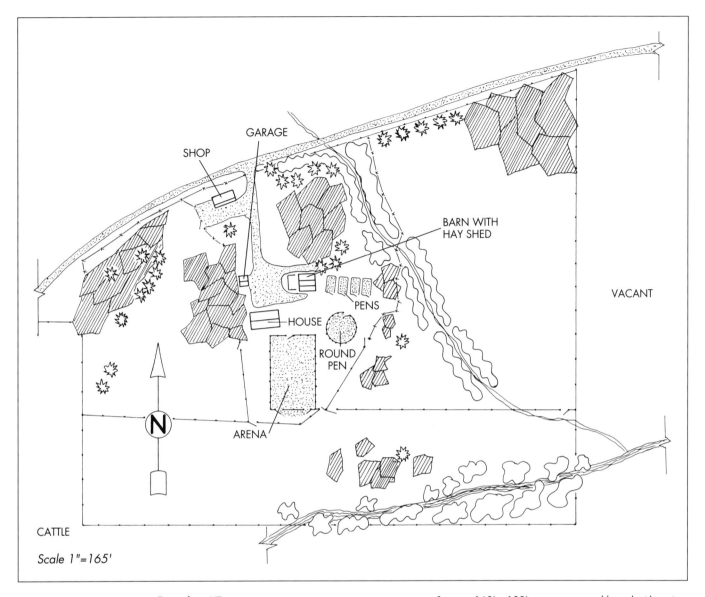

GARAGE

SHOP

BARN WITH
HAY SHED

VACANT

PENS

HOUSE

ROUND
PEN

ARENA

N

CATTLE

Scale 1"=165'

Drawing #7

AFTER: 20 Acres, ranch country.

House: 35' x 60' log home with approximately five acres of non-pasture area. House location to obtain maximum wind protection from rocks.

Garage: 24' x 24'

Barn: (See Drawing #13 for detail.) Added tack room (see Drawing #18 for detail) and stalls inside; added 16' x 48' hay storage area.

Shop: 48' x 100', also accommodates tractor storage.

Fencing:
Cross fences of various types: metal posts with smooth wire and/or electric, buck fence where metal posts can not be driven due to rocky ground. Woven wire on the inside of existing smooth wire fence on the east and southeast fences to keep out neighbors' cattle. Also some post and board, and metal panels used for training and holding pens.

Arena: 160' x 100' square post and board with native soil (decomposed granite) footing. Location was dictated by the presence of rocks and the lay of the land. Excavation into hillside was required leaving 16-foot banks on the east and west sides. Pasture fences skirt these banks to prevent erosion from horse traffic.

Pens:
Round pen, 65' diameter round post and board with sand footing. Four metal panel pens 16' x 32'.

Pastures:
Utilized primarily for exercise as fragile mountain grasses afford limited grazing. South—approximately 6 acres, provides water from creek. West—approximately 3 acres. East—approximately 6 acres, provides seasonal water from spring.

Landscaping:
Wildlife area: Entire central area (between east and west pastures) is off limits to free horses, so is a protected habitat for wildlife.

Privacy Border: Trees planted along the fence between the spring and rocks provide visual screen from road.

building project or improvement.

Once you have chosen a builder, you will probably sign a contract guaranteeing that the work will be done as promised and that you will pay the agreed-upon price. Contract requirements vary from state to state so be sure you are knowledgeable of the pertinent laws. Contracts should be written in simple terms and should contain clauses that cover:

● A list of the attached contract documents:

Agreement. Brief statement of project and price.

General conditions. Definition of responsibilities of all parties to the contract.

The drawings. The plan or blueprints.

Technical specifications. Specific definitions of number, type, quality, and/or brand of materials and products to be used.

Supplement. Modifications to fit specific needs of custom project such as insurance, bonding, etc.

● Statement of project, price, and responsibilities.

● An outline of how subcontracts and separate (but related) contracts are to be handled.

● Procedure of how plan changes are to be handled.

● Procedure of how disputes are to be handled.

● Statement of completion date.

● Outline of how and when payments are to be made with interest charged for payments not made on time.

● Statement of insurance requirements and safety responsibilities.

● Procedure to accommodate price, material, or completion date changes.

POLE BARN SPECIFICATIONS LIST

Building Dimensions

Length _____ ft. *(to outside of wall framing)*

Width _____ ft. *(to outside of wall framing)*

Height inside _____ ft.

Sidewalls

POLES:

Treated with penta 8-lb. retention? _____

_____ in. x _____ in. or _____ in. diameter

Length _____ ft.; In ground _____ ft.

Spacing open side _____ ft.; closed side _____ ft.

CONCRETE PAD:

_____ in. diameter; _____ in. deep

SPLASH BOARDS:

Treated with penta 6-lb. retention? _____

Qty. _____ 2 ft. x 6 ft.; CEM _____ in. high

(Or concrete foundation specs _____)

WALL GIRT:

_____ ft. x _____ ft. spaced _____ in. apart

WIND AND CORNER BRACING: _____

_____

RAFTER PLATE:

Open side _____ ft.; Closed side _____ ft.

Connection truss to rafter plate _____ ft.

SIDING MATERIAL:

Aluminum _____ Description: _____

_____

Steel _____ ; coating: standard 1¼ oz. or 2 oz.

Wood _____ Description: _____

Other _____

INSULATION:

Thickness _____ in.; Type _____

PLASTIC VAPOR BARRIER:_____ ml. thick

INSIDE SURFACE MATERIAL:_____

EAVES: open _____

VENTILATOR:

Type _____ ; Size_____

Roof
Designed and certified to withstand 30 psf? _____

TRUSS: *Sketch truss design*

Length _____ft.; Spacing_____ ft. OC;

Glue nail _____ ; Metal plate _____

CHORD:

Top _____ ; Bottom _____ ; Diagonals _____

ROOF GIRTS:

(Purlins)_____ ft. x_____ ft. spaced_____ ft. OC

ROOFING:

Aluminum _____

Steel _____ ; coating _____

Other _____

*All end laps not less than 6 in.?*_____

Number of skylight panels _____

CEILING MATERIAL:_____

INSULATION:

Thickness_____ in.; Type _____

VAPOR BARRIER:_____ ml. thick

FLASHINGS: _____

GUTTERS: _____

DOWNSPOUTS: _____

OVERHANGS: _____

Windows
NUMBER:_____

SIZES: _____

TYPE: _____

Doors
NUMBER: _____

SIZES: _____

TYPE:_____

Floor
FLOORING:

Type _____Thickness _____

Reinforcement _____

Bracing
Sketch type with nail or bolt pattern
Is brace included on each post? _____

Hardware
DOOR:

Type _____ ; Track length_____ ft.

GUTTERS:_____

DOWNSPOUTS: _____

HEATER?: _____

FANS?: _____

Utilities:
ELECTRICAL CIRCUITS:

Number_____ ; Wire size _____

LIGHT SOCKETS: Number _____

OUTLETS: Number_____

WATER OUTLETS: Number _____

DRAINS: Number _____

● Outline guarantees and procedure for dealing with defective work.

● Procedure to terminate the contract.

Rules and Regulations

You have certain obligations as a landowner and a horseowner. As far as protecting other people, animals, and property from your horses, one of your main obligations is to install and maintain a good fence. Look ahead to anticipate what potential problems are likely to occur, since you could be held liable for all results of your negligence. For example, a loose horse could kill or injure a human or animal, could cause a human or animal to injure themselves, could infect other horses with disease, could become involved in an unwanted breeding, or could inflict damage to buildings, fields, plants, or yards. If you have a stallion, you must be absolutely sure the animal is under control at all times — a stallion is potentially too dangerous to risk having it run loose.

Usually fence laws are covered by a state statute. Situations not specifically covered by law may be determined by court decisions or by conjecture. Often state laws say that a person has the duty to fence animals in and his neighbors do not have the responsibility to fence them out. Some open range states require landowners to fence out livestock.

If you share a division fence with a neighbor you usually share the cost of installation. However, adjoining owners usually cannot legally force each other to build a fence out of particular materials. Once the fence is built, you maintain the half of the fence to your right as you stand on your property at the midpoint of the fence facing the division line. It may be wise to have fence agreements with your neighbor in writing.

No matter who does the work on your land, buildings, or fences, it is ultimately your responsibility to comply with all legal codes and regulations. Check the local zoning regulations. These restrictions are designed to control the growth and development of communities by establishing particular areas for certain uses such as residential, commercial, industrial, agricultural. These laws define the type of building you can construct and the type of activity that can take place on your land. Zoning laws may also dictate building height and size, property size, legal distance from road or neighbors, and appearance of facilities. You will need to get approval from the zoning committee for your plans.

Your buildings must be constructed in accordance with the local building code. This a group of regulations or construction standards concerning structural soundness, enacted by law. They cover such topics as height and area restrictions, room size requirements, required method for design, minimum design loads for wind and snow, building materials, electrical requirements, plumbing and septic requirements, heating, ventilating, air conditioning, and sprinkler specifications. There are four major, independently developed building codes in the United States — the Standard Building Code, the Uniform Building Code, the Basic Building Code, and the National Building Code — and your local building official will advise you which one you must follow. You will need to submit plans for approval and get a building permit. The local building inspector may visit the construction site to ensure that the building is being constructed according to the approved plan.

Finally, you must check with the health department to be sure you are designing your farm to comply with public health requirements, pollution criteria, and pest control standards.

CHAPTER SEVEN

Barn Construction

A BARN SHOULD PROVIDE A SAFE, COMFORTABLE, and healthy home for your horses. Keep this in mind as you design it. Decide what features you need in your barn and then select a plan with those components or choose a builder that can create a custom barn for you. A common mistake is to accept a package deal that may not have everything you wish yet includes features that you may never use.

The site for your barn should be properly prepared. The barn floor should be 8 to 12 inches above ground level, and located on well-drained soil. The addition of 6 inches of crushed rock covered by tamped clay is a traditional barn favorite if the existing subsoil is well drained. Poorly drained soils should be excavated to between 3 and 10 feet, and several feet of large rock should be laid at the base of the excavation. Crushed rock of decreasing sizes should follow in layers, leaving about one foot for the barn's topsoil. This can be tamped clay or a mixture of three parts clay to one part sand.

If the soil is too soft, loose, or weak, and its bearing capacity is inadequate for the footings (support structures) of the foundation, the design engineer of the barn will have to make adjustments in the location or depth of the footings, or the cement forms for the footings. The barn should have a strong foundation made of either brick, concrete, or pressure-treated wood.

There should be plenty of windows or doors to let the sun and air in but keep the cold wind, rain, and snow out. Design your barn so that it can be warm in the winter but cool in the summer. A temperature range of 45 to 75 degrees Fahrenheit is best for horses, with 55 degrees being the ideal. A humidity of 50 to 75 percent is good with 60 percent optimum: however, it is better to be a little too dry than damp. Horses need adequate ventilation but cannot take cold drafts. Think of your fire plan as you design your facility. Design some lockable areas, such as a tack room and office, for security and insurance purposes.

Because horses roll, kick, and sometimes buck while in their stalls, the structure must be very strong. In addition, all hardware, bolts, doors, handles, latches, locks, and hinges must be heavy-duty to withstand horse use. Stalls, alleyways, and doorways must be safe with no protruding parts or narrow openings. Heavy traffic areas should be well sloped and drained and have a

Barns.

an inside aisle isn't essential, so many southern barns are simply single rows of stalls that open to outside pens or runs. Cold climates require inside access to the stalls. A very simple and popular style consists of two rows of stalls that face each other, separated by an inside aisle.

Closed barns are either uninsulated, insulated, or insulated and heated. Heated barns are expensive and an unnatural environment for horses, tending to result in more respiratory illnesses. Insulation is an air-filled or material-filled space between the inner and outer walls. It can include blanket, rigid, sprayed-on, and foamed-in-place products. Blanket type is usually foil or paper-backed and comes in rolls. Rigid insulation is usually a sheet of pressed fibers, often with a vinyl-coated side that can serve as the inside surface of the building. Spray-on cellulose fibers, though inexpensive, can absorb moisture in a humid climate and cause condensation and corrosion problems. Spray-on foam plastic can be useful for both roofs and walls but, according to most building codes, must be covered. Insulation prevents condensation by keeping the temperature of the interior walls the same as the air inside the structure.

A vapor barrier prevents or minimizes the flow of water vapor into the walls. The vapor barrier should be located in or near the warm side of the wall such as between the insulation and the inside wall. Special paints or waterproof membranes such as laminated paper, plastic, Kraft-backed aluminum foil, or foil-backed gypsum board are used for vapor barriers.

The overall shape of your barn is usually decided by the roof type and whether you plan to have a loft in your barn. Storing hay or bedding in a loft over the stalls does provide some insulation for cold climates but is such a potential fire hazard that it is strongly recommended to locate your hay storage in a building separate from the stable.

The gable roof is very popular and allows great flexibility in layout. The shed

protective, non-slip surface that is appropriate for the use and the locale.

The barn should be located with good access to electricity and water and situated so that there is room for future addition if desired. There should be convenient access from feed storage to the barn and from the barn to exercise and training areas. Many traditional designs and techniques have stood the test of time but new materials and innovations are worth considering.

Barn Types

There are many decisions for you to make when planning your barn. In warm climates,

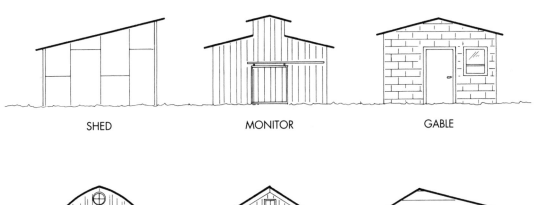

SHED   MONITOR   GABLE

GOTHIC   GAMBREL   OFFSET GABLE

Drawing #8
Roof Types.

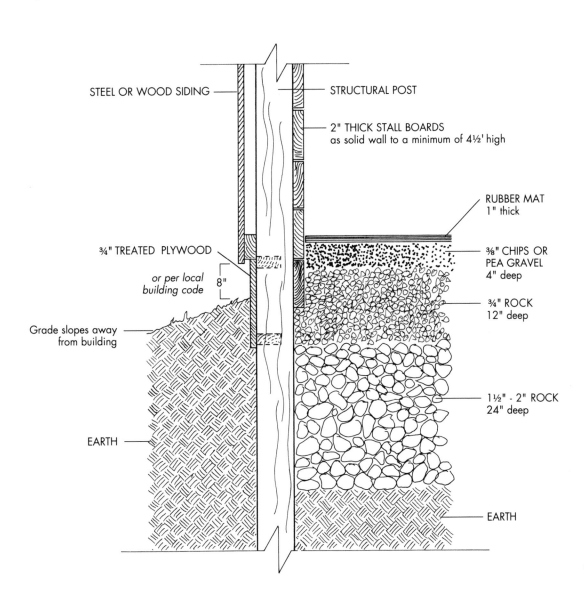

STEEL OR WOOD SIDING

STRUCTURAL POST

2" THICK STALL BOARDS
as solid wall to a minimum of 4½' high

RUBBER MAT
1" thick

¾" TREATED PLYWOOD

*or per local
building code*

8"

⅜" CHIPS OR
PEA GRAVEL
4" deep

¾" ROCK
12" deep

Grade slopes away
from building

1½" - 2" ROCK
24" deep

EARTH

EARTH

Drawing #9
**Ideal subsoil layering
for a barn.**

roof is often used for three-sided shelters or small stables or as an addition to an existing building that has a gable roof. The monitor is essentially two shed roofs with a gable in the middle. This is good for long rows of stalls. The area under the upper gable roof can be windows, vents, or clear panels.

South-facing, open-style stucco barn.
COURTESY OF PLEASANT VALLEY FARM

Horse barns are commonly of pole, frame, or masonry construction. Pole barns are quick, economical buildings. They usually consist of 6-to-8-inch-diameter pressure-treated posts set 3 to 6 feet below the ground with the bases fixed in concrete. The poles are set at from 8-to-16-foot intervals and have trusses attached to support the roof. Since the need for vertical support beams in the center of such a building is eliminated, the result is a clear inside span which makes for very flexible barn planning, the possibility of indoor riding spaces, and ease of expansion.

Frame or masonry barns require footings and foundation walls that extend out of the ground and attach to the barn walls. Where the outer walls of the building will be a trench is dug to below the frost line (the maximum depth the ground freezes in the winter) or according to the appropriate building code. Concrete footings are formed and poured in the bottom of the trench to trans-

fer the load of the structure to the soil. The foundation walls of concrete block or poured concrete sit on the footing and extend about 16 inches above the ground level.

Barn Materials

When choosing the materials for your barn walls and roof, consider cost, durability, maintenance, fire resistance, and aesthetics.

Walls

Metal (steel or aluminum) buildings are quick to put up, less expensive than wood, and require minimum maintenance. They can be noisy, however, during windy or rainy weather, they can be cold in the winter and hot in the summer, and although they are neat in appearance they may not be thought as aesthetically attractive as other choices. In a metal building the stalls must be lined with planks at least 6 feet up from the ground to prevent damage to the metal siding by the horse.

Wood buildings are traditional, attractive, and provide good insulation. They are expensive, however, can require more labor for construction, and need frequent maintenance. Horses chew wood, so surfaces must be treated. Wood buildings are a fire hazard and should be built of fire-retardant materials and have fireproof lining where possible. Wood or fiberboard ship lap siding or vertical boards are often used on the exterior.

Though treated plywood may not be as attractive as planks, it is very strong. It is a manufactured wood product with a solid or veneer core covered by thin layers (plies) of wood. Each ply is placed with its grain at right angles to the next ply, and all of the layers are laminated together with glue. Plywood therefore has a much higher strength-to-weight ratio than lumber has. Three-quarter-inch plywood is usually an

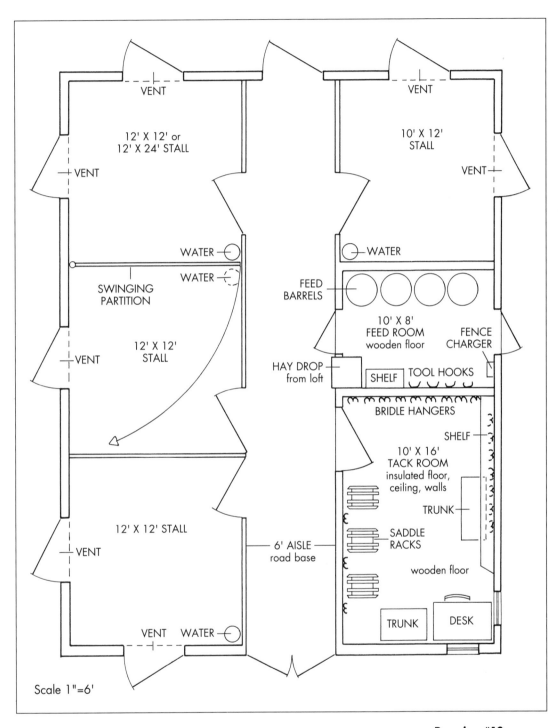

VENT

12' X 12' or
12' X 24' STALL

VENT

WATER

WATER

SWINGING
PARTITION

12' X 12'
STALL

VENT

VENT

12' X 12' STALL

VENT WATER

Scale 1"=6'

VENT

10' X 12'
STALL

VENT

WATER

FEED
BARRELS

10' X 8'
FEED ROOM
wooden floor

FENCE
CHARGER

HAY DROP
from loft

SHELF

TOOL HOOKS

BRIDLE HANGERS

SHELF

10' X 16'
TACK ROOM
insulated floor,
ceiling, walls

TRUNK

6' AISLE
road base

SADDLE
RACKS

wooden floor

TRUNK

DESK

Drawing #10
Barn plan from
Layout Drawing #5.
Vents are 5 feet high
from floor and are
covered with heavy
mesh panels.

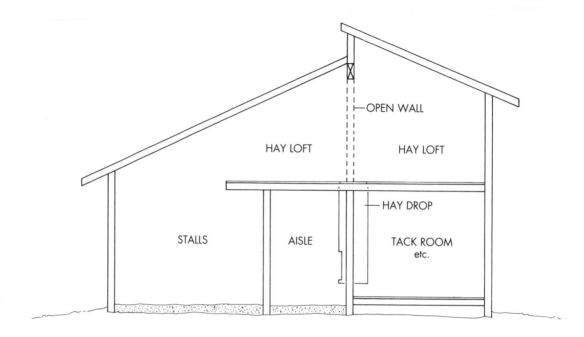

Drawing #11
Profile of barn in
Drawing #10. Note
loft for hay storage
and hay chute to
lower level. Trans-
lucent panels in
south facing short
wall formed at junc-
tion of two roofs.

OPEN WALL

HAY LOFT   HAY LOFT

HAY DROP

STALLS   AISLE   TACK ROOM
                               etc.

adequate replacement for 2-inch planks for the exterior of a building. Plywood won't warp, split, or shrink the ways boards will.

Masonry-type buildings include brick, concrete block, poured cement, and stone. These buildings are cool in warm climates but can be cold and damp in cold climates unless properly engineered. Generally there is a high cost in both labor and materials for construction. Because such structures are virtually fireproof, they often qualify for lower insurance rates. Masonry-type buildings must usually be mechanically ventilated, however, in order to produce a satisfactory environment for horses. The use of a vapor barrier often increases condensation in these buildings.

Roofs

Metal roofs are quick to install, relatively inexpensive, and maintenance free. They can be very noisy, however, may leak around the fasteners, are prime candidates for condensation problems, and can make the barn very hot during sunny weather. Zinc-coated steel or sheet iron is durable and economical but not particularly attractive and is hot unless insulated and vented. Aluminum sheets with baked-on colors are more attractive but may not be as strong as steel roofing. Translucent fiberglass panels can easily be incorporated in metal roofs for light. Snow slides off metal roofs, often in huge sheets and without warning. Plan doorways and runs with this in mind.

Redwood or cedar shakes make very attractive roofs that can last fifty years or more. Shake roofs are warm even without insulation, cool in summer, and condensation is not a problem. They are expensive and the labor cost to lay wood shingles is very high. They are considered a fire hazard and may affect your insurance rates. Shakes should be used only on roofs with a ¼ pitch so that water drains off and the shingles dry out quickly.

Asbestos cement shingles are fireproof, attractive, and have a fifteen- to twenty-year life span. They are very expensive, however, and difficult to install. They must be nailed individually to solid sheeting and they are very prone to wind and hail damage.

Asphalt roofing, both shingles and rolled roofing, will last twenty years if properly installed. It is relatively inexpensive, comes in a range of colors, is one of the least complicated roofings to install, and is quieter

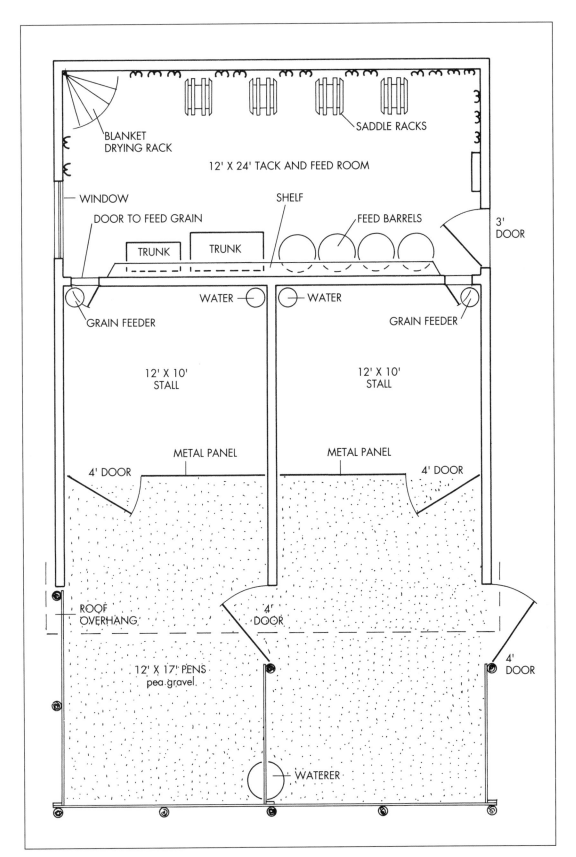

BLANKET
DRYING RACK

SADDLE RACKS

12' X 24' TACK AND FEED ROOM

WINDOW

SHELF

DOOR TO FEED GRAIN

FEED BARRELS

3'
DOOR

TRUNK   TRUNK

WATER   WATER

GRAIN FEEDER   GRAIN FEEDER

12' X 10'
STALL

12' X 10'
STALL

METAL PANEL   METAL PANEL

4' DOOR   4' DOOR

4'
DOOR

ROOF
OVERHANG

4'
DOOR

12' X 17' PENS
pea gravel.

WATERER

Drawing #12
**Stable 1 from
Layout Drawing #3.**

FLOORING CHOICES

TYPE	INITIAL COST	LONGEVITY	CUSHION	DRAINAGE	ODOR
DIRT	low	1-2 years in stall	good	forms mud	yes
CLAY	low	1-2 years in stall	good	slow drying	yes
SAND	low	1-2 years	very good	good	mild
GRAVEL	low	5 years	shifty; rough	good	mild
ROAD BASE (Gravel/Limestone/Dirt Mix)	low	1-2 years	OK	good	mild
CONCRETE	medium	permanent	none	needs drain or slope	no
BRICKS	high	permanent	none	needs drain or slope	no
RUBBER	high	10 years or more	good	needs base or slope	no
WOOD	high	2-3 years	OK	needs spaces	yes
ASPHALT	medium	5 years or more	OK	needs slope or drain	no
I/O CARPET	high	5 years	OK	needs slope or drain	mild

than steel roofing. It is more susceptible to damage by wind, cold, ice buildup, and extreme heat than steel roofing.

Felt roofing paper provides a vapor barrier for a roof and is useful as an under layer but not as an exterior roofing, as it simply will not last.

Miscellaneous Exterior Features

To handle water from the roof during a rain, you may wish to consider the inclusion of gutters, downspouts, and rones (concrete splash pads).

To keep entryways from becoming muddy when snow slides off the roof, you may wish to attach overhangs to the roof to shelter the doorways. If you plan to attach exercise runs to the barn, locate them so that rain and snow accumulation from the roof does not empty into the runs.

To prevent fire by lightning strike, you should include a properly installed lightning conductor to the most prominent roof. (See chapter 15 on fire prevention for more details.)

Flooring

You will probably end up with three or more types of flooring in your barn. Stalls, aisles, the tack room, the feed room, the wash rack and the office all have different flooring requirements. Things that you need to consider for any floor are cushion, absorbency or drainage, cleaning convenience, traction, odor retention, moisture retention and, of course, that pesky item, cost.

Some floor areas will need to be sloped

EASE TO CLEAN	SANITATION MAINTENANCE	TRACTION	SUITABILITY	OTHER COMMENTS
difficult	difficult	good	stalls	frequent maintenance and replacement
difficult	difficult	good	stalls	frequent maintenance and replacement
difficult	difficult	good	none	sand colic; sand cracks
difficult	difficult	OK	aisles	use pea gravel
difficult	difficult	good	stalls; aisles	can be dusty
easy	easy	good if texturized	wash rack; feed room; shoeing	
easy	easy	OK, but slick when wet	aisles; tack room; office	attractive
easy	easy	good if texturized	stalls; wash rack; shoeing	OK over concrete
OK	difficult	slippery when wet	tack room; office	
OK	possible	good	aisles	
OK	possible	good	aisles; tack room office	

to encourage surface drainage but not more than 1 inch in 5 feet. Other areas should provide level spots for work such as the farrier's slab. Too much slope or irregular footing here makes accurate work difficult and is uncomfortable for both the farrier and the horse. Most floors must provide adequate traction, but especially those that get wet such as stalls, wash racks, and grooming areas. When choosing your floorings, remember that they must be raked and/or swept regularly and occasionally washed and disinfected.

As you look over the information in the flooring chart and begin talking with dealers, realize that some floorings are made to be used in conjunction with others, so you may end up choosing two floorings for one space. For example, rubber mats over tamped road base for stalls or indoor/outdoor carpet over wood for the tack room.

Lighting

Incorporate a combination of natural and artificial light in your stable plans. Natural light should include some direct sunlight, a source of vitamin D. Sunlight is also one of the best sanitizing agents and is essential for keeping a barn from developing a rank odor. Sunlight also adds warmth, however, and although it may be greatly appreciated in the winter, extra heat can become a big problem in the summer. Natural light can enter the barn through windows, large sliding doors, and through panels and vents in the roof.

Artificial lighting can be of three types: incandescent, fluorescent, or mercury vapor.

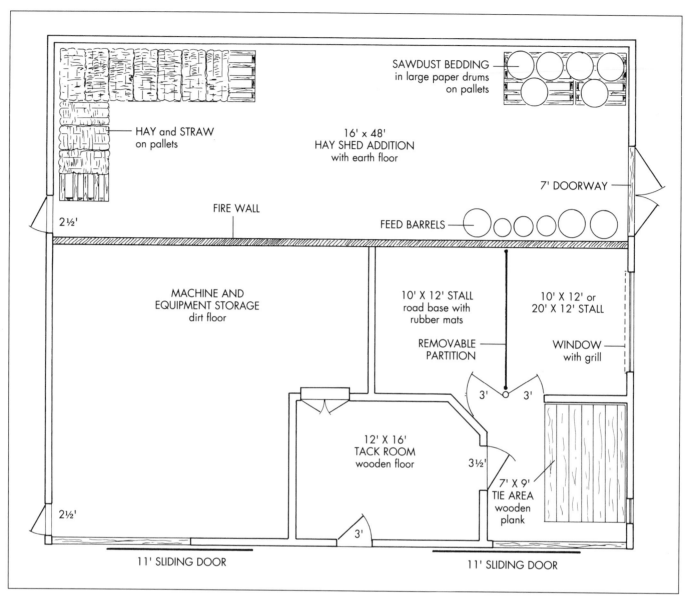

SAWDUST BEDDING
in large paper drums
on pallets

HAY and STRAW
on pallets

16' x 48'
HAY SHED ADDITION
with earth floor

7' DOORWAY

FIRE WALL

2½'

FEED BARRELS

**MACHINE AND
EQUIPMENT STORAGE**
dirt floor

10' X 12' STALL
road base with
rubber mats

10' X 12' or
20' X 12' STALL

**REMOVABLE
PARTITION**

WINDOW
with grill

3'   3'

12' X 16'
TACK ROOM
wooden floor

3½'

7' X 9'
TIE AREA
wooden
plank

2½'

3'

11' SLIDING DOOR

11' SLIDING DOOR

Drawing #13
Barn plan from
Layout Drawing #7.

Common household bulbs (incandescent) are satisfactory for stalls, tack rooms, and so on. Fluorescent light, though bright, is not practical for high ceilings because it diffuses quickly, so it is best in a work area where you may need concentrated lighting. To get illumination equivalent to a 150-watt incandescent bulb, you will need to use a 840-watt fluorescent bulb.

Mercury vapor lights, which are several times brighter than fluorescent, are best used in very high ceilings, such as indoor riding arenas, as they need at least sixteen feet to diffuse evenly.

All lights, switches, and receptacles must be out of the reach of your horse's curious lips. Bulbs in stalls should be fitted with wire baskets or recessed into the ceiling

with a panel over them. Depending on the receptacle and ceiling, a 100-watt incandescent bulb should illuminate a 12 x 12 box stall sufficiently for you to clean it. If you locate the lights above the stall dividers and toward one corner, you may have fewer dark spots in the stall and may be able to use flood lamps in swivel fixtures, providing they are well out of a rearing horse's reach.

All electrical wires should be threaded through metal or hard rubber conduit pipe to prevent rodents from chewing through the wires, one of the leading causes of stable fires. You should have a separate fuse box for the stable to prevent domestic overload. Check your local building code for the required number of receptacles for your structure but plan for more electrical outlets

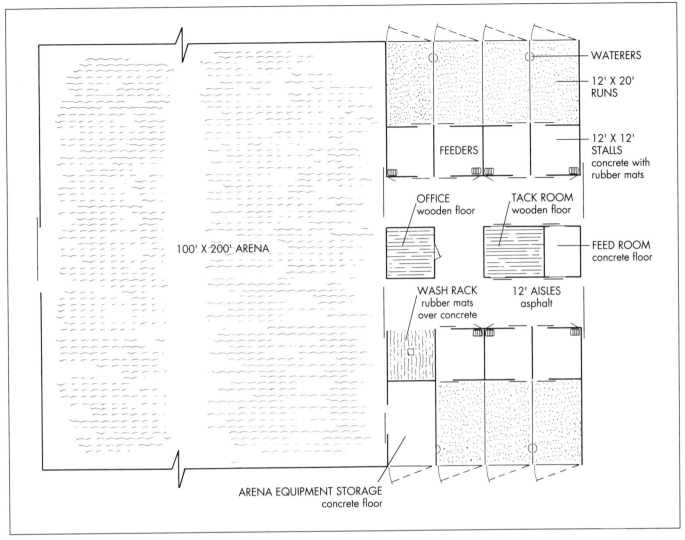

WATERERS

12' X 20'
RUNS

12' X 12'
STALLS
concrete with
rubber mats

FEEDERS

OFFICE
wooden floor

TACK ROOM
wooden floor

FEED ROOM
concrete floor

100' X 200' ARENA

WASH RACK
rubber mats
over concrete

12' AISLES
asphalt

ARENA EQUIPMENT STORAGE
concrete floor

Drawing #14
**Barn plan with
attached indoor
arena.**

than you think you will need. Consider that you may use any or all of the following: extra flood lamps, clippers, vacuum cleaner, heaters, battery chargers, electric waterers, hot plate, coffee pot, heat lamps, radio, refrigerator, and so on. Try to avoid using extension cords and never overload plugs. Install three-way switches wherever practical. The convenience of having switches in two locations for one light will ensure that lights are turned off when they should be.

Ventilation

A 1,000-pound horse releases two gallons of moisture into the air each day through respiration. A four-horse barn must thus deal with more than eight gallons of water vapor per day, not counting the additional moisture created by the evaporation of urine and manure.

Warm air can hold water in suspension but as the temperature drops, it loses as much as one half its water-carrying capacity. The moist air rising from a horse's body condenses on the underside of uninsulated roofs, causing dripping, dampness, and sometimes ice formation. Damp air contributes to respiratory ailments, stiffness, bacterial growth, and fungal growth.

Ventilation, the movement of air through the barn, regulates the temperature and humidity. The goal is to make the air in the stalls and the aisles fresh. Check your design to assure that ammonia fumes from manure and bedding decomposition can move up and out of the stalls. Also plan your structure to prevent condensation which occurs when the difference between indoor and outdoor temperature is too great. Condensation does not occur as readily in moving air as it does in stagnant air.

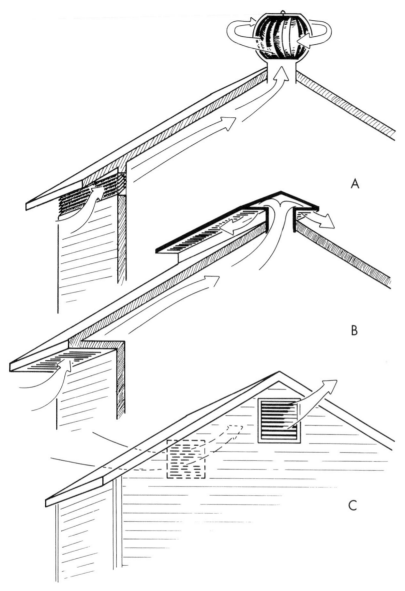

Drawing #15
Ventilation.

A. Upper wall louvres to turbine vent.

B. Soffit louvres to continuous ridge vent.

C. Gable louvres.

Ventilation can be provided by natural and/or mechanical means. One of the easiest methods is to open windows, barn doors, and the top of stall doors. Of course, you must beware of drafts as horses are very susceptible to chill. Stall windows should be outfitted with a metal grill with spaces no more than three inches apart to protect the glass from horses. Metal window frames are better than wood as they are fireproof. If a window is set high in the stall, it can be of the type that opens inward and is hinged at the bottom.

Doors should be a minimum of 8 feet high and 4 feet 6 inches wide. If a door is to be left open, it should have a latch or hook to keep it there. Sliding doors should be outfitted with proper tracks, and with bumpers to prevent them from rolling off the tracks.

Various vents can be placed in the barn to allow foul air to escape. All sorts of variations of louvre boards can be implemented in the roof ridge or under the eaves in the soffitt. A spinning cupola ventilator can be added on the roof as a supplement to ridge vents. Other non-mechanical means of ventilation include adjustable vents 6 feet from the stall floor or vents at the top of the back wall where it joins the roof.

Mechanical ventilation forces air in or out of the barn, usually with electrical fans or blowers. When figuring either for a pressure or an exhaust system, figure one square foot of vent or inlet space for every 750 cubic feet per minute fan capacity. More than this will create a draft, and less will simply not ventilate adequately.

CHAPTER EIGHT

Inside the Barn

A HORSE BARN CAN BE VERY SIMPLE OR VERY ELABORATE. The most expensive barn is not necessarily the best environment for horses; nor does it ensure efficiency and convenience of routine management. Decide what the main purpose of the barn will be, where you will be spending the majority of your time, what work areas need to be roomy and well equipped, and how many stalls you will realistically need. Spend time with a pencil and some graph paper trying out your ideas.

The Stalls

Whether you are designing a new barn, remodeling an old one, or would like to make a few changes in your horse's present stall, keep your horse's comfort and safety foremost in your mind. This includes considering durability and sanitation as well as convenience of cleaning and feeding routines.

Stall Types

Horses are usually kept either in box stalls where they can move about freely or in tie stalls where they remain tied. Tie stalls are only somewhat larger than the traveling space in a conventional horse trailer. Tie stalls are OK for separating horses at feeding time, but they are unsuitable for longterm horse housing. A horse simply cannot get quality rest in a 5½-foot-by-10-foot enclosure. Some horses are hesitant to lie down at all when their heads are tied to the manger. In order for a horse to lie down comfortably, the lead rope must be long, which then presents a potential hazard — the horse may get tangled in the rope. Also, tie stalls prohibit exercise. A horse in a tie stall shifts his weight from quarter to quarter in a response to fatigue and lack of blood flow. Box stalls, which are roomier and more comfortable, are today's choice for the horse who is kept in for training, showing, rehabilitation, or foaling.

Box Stall Size

A 12-foot-by-12-foot box stall is appropriate for most horses. The 144-square-foot area that such a stall provides seems to be optimal when considering the horse's comfort and the stall's maintenance. This size offers

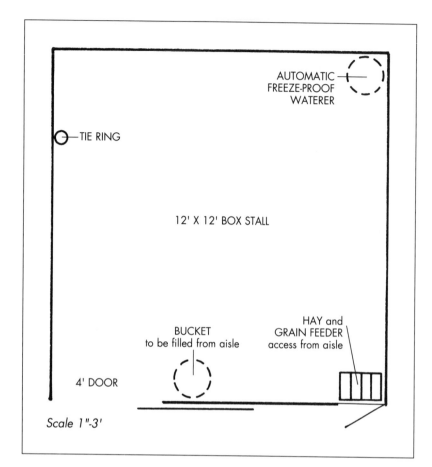

AUTOMATIC FREEZE-PROOF WATERER

TIE RING

12' X 12' BOX STALL

BUCKET
to be filled from aisle

HAY and GRAIN FEEDER
access from aisle

4' DOOR

Scale 1"-3'

**Drawing #16
Box stall with two
watering options.**

enough room so that a horse can confine his defecation and urination to a certain portion of the floor and still have plenty of clean space for eating and resting. A smaller stall can result in a horse inadvertently churning manure into the bedding every time he moves. This means more labor for stall cleaning, greater waste of feed and bedding, and a greater potential for parasite reinfestation. Stalls much larger than 12-feet-by-12-feet may allow more space between defecation, feed, and rest areas, but the cost of building space and bedding may be prohibitive.

An important aspect to consider when determining stall size is the behavior of each individual horse. It may be to your advantage when planning a barn to have a few 10-by-10 stalls for small horses or ponies or for the horses who defecate anywhere and everywhere and mince their manure into the bedding regardless of the size of their living

quarters. In addition, you may wish to consider a few oversized stalls for the very large horse, for the horse that rolls frequently, or for the rehabilitating horse.

The foaling or nursing mare requires a double stall. You can plan your barn so that two of the regular-sized stalls have a removable partition between them. This will allow you to use the space either for two single horses or for one mare and foal. You may wish to locate the foaling stall in a position in the barn that allows observation, such as from a tack room. You also may wish to plan other specialized stalls, such as for stallions or for isolating incoming or sick horses.

Stall Construction

Stall walls are often made of wood, metal, or cement block. They must be specially designed to withstand kicking, rubbing, wood chewing, and the rotting and corrosive effects of urine and manure. Be sure the interior, especially, is smooth, free of any projections, rugged, and has no exposed wood edges. Even the end of a bolt protruding to the inside of the stall can be a potential hazard, for as you probably know, if there is a way to get hurt, a horse will ultimately find it.

Barns usually are designed with either cement or pressure-treated wood foundations. The foundation, set in trenches at wall lines, carries the weight of the building and mediates between the soil and moisture outside the barn and the flooring and bedding inside the barn. Some sort of foundation should also be used under the walls between stalls and along the alleyway. If these interior walls support no load other than their own weight, their foundations can be quite shallow.

The lower 4 to 5 feet of a stall wall is often a solid material such as wood or cement block. If lumber is used, it should be a full 2 inches thick to withstand kicking, and this wall lining should have no spaces for legs to get caught in when the horse rolls. Unless lumber is ordered Rough Sawn (RS), which is how it comes from the sawmill, it is dressed (surfaced with a planing machine) so that its actual dimensions are less than the "nominal dimension" by which it is ordered. A Rough Sawn 2 x 8 is a full 2 inches thick whereas a dressed 2 x 8 is only 1½ inches thick.

Most lumber for construction comes from softwoods, pine and fir being the most common. Fir is the stronger species but is harder to drive nails into, splits more easily, and will splinter, posing a possible hazard to horses. Two-inch pine boards would be fine for the stalls. Choose boards that are straight-grained, free of large knots (where breaking usually occurs), and not warped. White oak, a hardwood, is more expensive than pine but is a very strong wood and has the great advantage of a chemistry that naturally deters wood-chewing by horses. Plywood, which comes in 4-by-8-foot sheets, can also be used for the lower portion of stall walls if sufficient backup framing is provided.

The top portion of the front stall wall is generally made of mesh, pipe, or bars to allow the horse to see out and to ensure proper ventilation. Spaces larger than two inches between bars or in the mesh can be dangerous. Nibbling horses can get their teeth or jaw caught and inflict serious damage to themselves. Because of confined horses' tendency to mouth and play with stall fixtures, all exposed wood edges should be covered with a chemical chewing deterrent and/or metal strips.

To discourage fighting between stalls, the dividing partition can be solid and a minimum of 8 feet high. The ceiling of a stall should be somewhere between 8 and 12 feet high.

Stall doors are usually hinged, sliding, or a combination of the two. Hinged doors that open outward can block the alley, making it awkward to take a horse in or out unless the aisle is absolutely clear. Hinged doors that open inward crowd both the horse and

Because protrusions increase likelihood of injury, this stall latch has been recessed for safety.

All wooden edges in a stall should be covered with metal strips.

the handler and can cause either to get wedged if the door should happen to catch on horse or handler on the way out. Dutch doors, those comprised of two half-doors, are traditionally hinged to open outward. The top can have a wire mesh panel if the horse lunges at passing horses or handlers. Otherwise it can be fastened in its open position so that the horse can put his head over the lower door and see other horses and perhaps get some fresh air and sunlight. Sometimes a horse will develop a habit of hanging on the lower door, which is damaging to the hinges.

Sliding doors are convenient and space-efficient because when open they fit closely along the front of the stall wall. It is important to be sure the sliding door has a roller at its base to keep the bottom of the door from being pushed outward when a horse rolls against it. Sliding doors are also available with a dutch-door-type feature. A drop panel in the top of the sliding door will allow the horse to put its head out. Whatever type of door is chosen, it should be a minimum of 4½ feet wide and at least 8 feet tall. There should be two latches for all stall doors, one at the bottom of the door, out of the horse's reach.

Aluminum scoop shovel for cleaning stall with rubber mats.

Stall flooring

Since a horse can produce up to 40 pounds of manure and 10 gallons of urine daily, a good deal of thought must be given to the stall flooring and bedding. Flooring choices were discussed in the previous chapter. As you make your decision, compare the initial cost of installation versus the durability and longevity of each product. Weigh that along with the margin of safety and comfort for the horse provided by each type of flooring. And finally, be sure to consider what type of bedding you plan to use as some flooring/bedding combinations can be undesirable.

The tamped clay floor is a traditional favorite for stalls because it provides cushion, good traction, and is warm and quiet. However, clay does not percolate well and stall floors must be sloped to allow drainage. In addition, urine pools soon become potholes of enormous proportions requiring that the clay floor be leveled routinely and overhauled annually. Two to four inches of the original 6 to 12 inches are removed each year and replaced with fresh clay and retamped. Pure clay may be difficult to buy in some areas so if you decide to go with clay, you can extend the life of the stall by allowing it to rest periodically until dry. If you have at least one extra stall in your barn, rotating horses will allow one stall to be empty at all times.

Mixtures of clay and sand or crushed rock may be more readily available than pure clay and will have improved drainage while retaining most of the clay floor's desirable features. Road-base in your locale may be such a mixture — a blend of crushed lime-stone and clay. While most of these blends may result in better sanitation, however, they are more unstable under a horse's movement, can become mixed in with the bedding or ingested with the feed, or can be a pawing horse's delight.

Concrete makes a permanent, low-maintenance floor that is fairly easy to sanitize. However, it requires very deep bedding because it is hard, cold, and abrasive. Concrete floors must be designed with proper slope for drainage and should be finished rough or scored to ensure good traction.

Asphalt (blacktop) is a relatively permanent material that has some give, is non-slip, and is not as cold as concrete. However, it is abrasive, requires deep bedding, and is initially one of the most costly choices.

"Rubber" stall mats can be thought of as part flooring, part bedding. The mats are usually a combination of rubber, clay, nylon, and rayon. They act as an intermediary between the soil and the bedding. In this way they prevent horses from ingesting dirt or sand with feed they eat off the stall floor. They have superior cushioning so they are comfortable, can be easily sanitized, make stall cleaning easy, decrease dust so horses stay cleaner, and decrease the amount of bedding required by up to half. They also prevent a pawing horse from digging holes in the stall. Stall mats are initially expensive, however, and if they are not properly texturized, they can be slippery.

Wood, an old-time favorite for tie stalls, is really not appropriate for box stalls. Although it is warmer than concrete and fairly durable if appropriate wood is used, it can be slippery, is difficult to sanitize and deodorize, and can be noisy under a nervous horse.

Bedding

What type of bedding do you prefer and is it readily available in your area? Availability greatly affects price. In the bedding chart on page 68, price estimates are general. You may find a certain bedding to be very economical in your area depending if you are in a timber area or farmland. Bedding should be easy to handle, absorbent, and encourage a horse to lie down. Deciding ahead of time what type of bedding you will use will probably affect your choice of stall flooring.

The bedding with the highest water-absorbing capacity is not necessarily the best bet. An extremely absorbent bedding sops up too much urine and the horse stands or lays in the soggy mess. On the other hand, a bedding with very little absorbency lets too much moisture pass to the flooring. The ideal bedding in most cases has an absorbency of between 2.0 and 3.0 and is free from dust, mold, and injurious substances.

Softwood products, such as pine sawdust, shavings, or chips are commonly used for bedding. Sawdust is made up of small particles from the sawing of logs into lumber. Sawdust from smaller saws, such as those in cabinet shops, may be too fine and dusty.

Shavings are very thin, small slices of wood produced by the planing or surfacing of lumber. Shavings are available at sawmills and cabinet shops. Chips are small coarse pieces of wood produced by the drilling, shaping, turning, or molding of lumber. Hardwood products are generally undesirable because of their poor absorbency and in some cases, such as with black walnut, a dangerous toxicity. Horses merely coming in contact with such shavings have experienced founder and death.

Wheat straw, because of its high glaze, does not become as slimy and sloppy as oat straw when wet and is less palatable to horses than oat straw so may be safer to use with the horse that overeats. Barley straw should be avoided because of the sharp, barbed awns that can become lodged in a horse's gums. Straw of any kind is very slippery on wood floors. All beddings, even peat moss, have the potential to be dusty. Be selective, don't purchase dusty bedding material and prevent respiratory problems in your horse.

Stall Details

Ventilation should have been considered in the overall plan for the barn, but certain stall features will help ensure that horses get adequate air flow without drafts. If part of the stall front is fitted with bars or mesh and a 2-foot-by-2-foot window is located on the back wall of the stall, the stall will be set up to take advantage of additional light and warmth from the sun or cool breezes, depending on the season and time of the day. All windows should be covered with heavy wire mesh or close-fitting bars on the stall side for safety. Translucent panels can be used in the roof to increase natural light and take advantage of heat from the sun. Unless the panels are fitted with shades, however, this may not be the best choice for the summer season in a very sunny or hot climate.

The feeding and watering features of a stall should be convenient for the manager and should allow the horse to eat and drink

BEDDING
(in order of absorbency)

BEDDING	LBS. OF WATER ABSORBED per lb. of bedding	COST	COMFORT	CLEANLINESS
PEAT MOSS	10.0	high	thick, soft bed	dusty; gets foul fast
PINE CHIPS	3.0	low	rough	can contain foreign objects
OAT STRAW	2.8	medium	good if not crushed	good
PINE SAWDUST	2.5	low	warm, soft bed	can contain foreign objects
WHEAT STRAW	2.2	high	good if not crushed	rarely dusty
BARLEY STRAW	2.1	high	short stem not elastic	often damp or dusty
PINE SHAVINGS	2.0	low	fluffy bed	can contain foreign objects
HARDWOOD CHIPS, SAWDUST, OR SHAVINGS	1.5	low	can be rough	can contain foreign objects
SAND	.2	low	soft but abrasive	dusty

Corner bucket with bracket.

Automatic waterers are convenient but must be inspected and cleaned daily.

easily and safely. Feeders should be located for easy filling and have a capacity for up to 20 pounds of hay. It is best for your horse's health if he can eat from a feeder at chest to wither height or lower. Feeders higher than the withers often cause dust or leaf particles to fall into the horse's nostrils or eyes and can cause respiratory problems and clogged tear ducts. Hay nets can be a good temporary way to feed hay but care must be taken to tie the hay net so that when it is empty it does not hang dangerously low. As an added safeguard, hay nets can be fitted with a breakaway fastener so that if the horse does become entangled he will not hang himself. Feeding hay on the ground is an excellent natural choice, as long as stalls are not bedded with sand and are kept scrupulously free of manure to prevent ingestion of parasite eggs.

Grain fed in shallow tubs on the ground is a natural way for horses to eat but many horses tip the tubs over and spill the grain. A good solution is to feed grain in the bottom tray of the hay rack. The tray should be lo-cated at about two-thirds the height of the horse or about 38 to 42 inches off the ground. Feeding grain in buckets is okay, but horses tend to chew and rub on buckets. They bang them and tip them over and will tend to gulp their grain.

Since a horse drinks from 8 to 12 gallons of water per day, it is difficult for a bucket system to work well unless you use a 3- to 5-gallon bucket and fill it two to three times per day. Waterers tend to stay cleaner if they are located away from the feeding area. They are safest if they are round and set in a corner. If they are square, the corners should be covered with a protective edging or, better yet, the entire waterer can be installed flush with the wall. If you have turn-out pens attached to each stall, you may wish to locate the waterers out there. Although heated, automatic waterers are convenient, they are expensive and you never really know if a horse is drinking, and if so, exactly how much. Also, with some models, horses learn how to keep the water running and flood

CONVENIENCE FACTORS	PALATABILITY	OTHER
when dry—light; when wet—very heavy	low	difficult to see manure; gets soggy; associated with thrush; good compost value
usually hauled bulk then must be shoveled	low	can be drying to hooves
light bales but slimy when wet	high	horse often gets "straw belly" from eating bedding
bags OK; bulk then must be shoveled	low	can pack and be drying to hooves, can ferment when wet and heat hooves
light bales; heavy when wet	low; but OK if eaten	food compost value but makes large manure pile
light bales; heavy when wet	awns irritate eyes and mouth; can cause colic	can cause colic
very light but large volume required	low	can be drying to hooves; hang in mane and tail
hauled bulk then must be shoveled	black walnut toxin can cause founder	not worth the chance of toxic effect if black walnut is present
heavy	inadvertently ingested with feed	may result in colic or hoof damage — sand particles can work their way into the bottom of the hoof wall

This "freeze-proof" cabinet provides a place to store medicines and grooming supplies.

their stalls. Be sure to check automatic waterers daily to be sure that they are clean and functioning properly.

Stalls are somewhat contrary to a horse's natural and preferred habitat, but a horse can learn to enjoy the comforts of a well-designed and well-managed stall. Make your horse's stall a safe and pleasant home. Doing so will help ensure that when you arrive to take him for a ride, your horse will be healthy, well-rested, and in a good state of mind.

The Tack Room

Since a tack room is often the hub of human activity in a barn, it should be roomy and well designed. If you are building a new barn or remodeling an old one, first consider the main purposes your tack room will serve.

Depending on the scope of the horse operation and the availability of other buildings and rooms, a tack room can end up being like a feed room or like an art gallery. Limiting the purposes for which it is in-tended will help to keep a tack room neat and functional. Do you need an area to organize your everyday working gear? To arrange your grooming and medical supplies? To safeguard your records? Do you need a place to store winter blankets? Show saddles and bridles? Do you require an area for saddle cleaning and repair? Does your tack room need to double as an office or trophy showcase?

A tack room should be well organized, dust free, insulated, well ventilated, dry but not hot, rodent free, secure, and provide plenty of storage space. If you can determine just how you will use the room you can estimate the optimal size and begin your floor plans. Within reason, make the room as large as possible. Rooms smaller than 8-by-10 feet (the size of a small box stall) seem crowded as soon as a few saddle racks and tack trunks are moved in. If you have from two to ten horses, try to allow between 100 and 200 square feet for a tack room in your barn blueprints. Using a piece of graph paper,

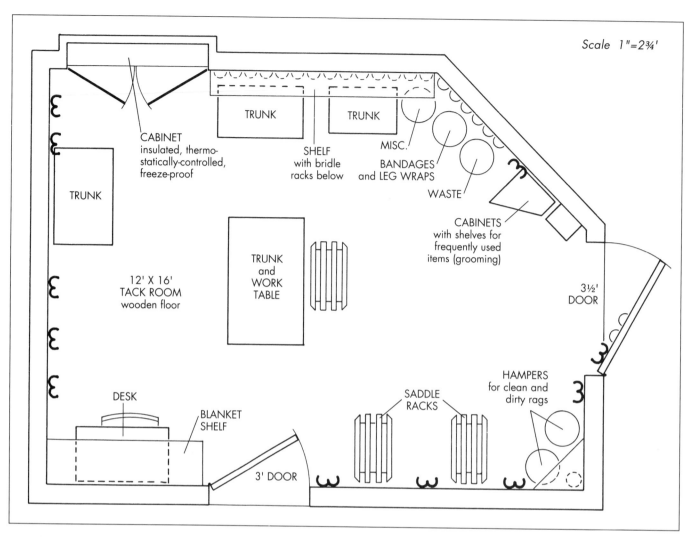

Scale 1"=2¾'

TRUNK

TRUNK

CABINET
insulated, thermo-
statically-controlled,
freeze-proof

SHELF
with bridle
racks below

MISC.

BANDAGES
and LEG WRAPS

WASTE

CABINETS
with shelves for
frequently used
items (grooming)

TRUNK

12' X 16'
TACK ROOM
wooden floor

TRUNK
and
WORK
TABLE

3½'
DOOR

DESK

BLANKET
SHELF

SADDLE
RACKS

HAMPERS
for clean and
dirty rags

3' DOOR

Drawing #17
Tack room plan from
Barn Drawing #7.

draw the proposed floor plan to scale noting the placement of saddle racks, tack trunks, desk, etc. A 10 foot by 22 foot (approximate inside dimensions) room can be substituted for two 12-by-12 foot box stalls in most barn plans and results in a very useful space. For a larger barn, two or more smaller tack rooms in several locations may be more convenient and efficient.

A tack room should be located very near the grooming and saddling area. Consideration should be given to air flow between the two locations so that dirt, hair, and sweepings from the grooming area are not automatically sucked into the tack room. The doorway between the grooming area and tack room should be at least 3½ feet wide to accommodate a person carrying a western saddle. All doors should be fitted with strong, durable locking devices to prevent theft and to satisfy insurance requirements.

Construction. Whether you do the construction yourself or hire someone else to do it, the work should be professional and comply with local building codes. After the tack room is framed, and before the walls are covered, the wires for the electrical outlets and lights need to be placed. Outlets should be plentiful, one every 6 feet. Locating the receptacles about 4½ feet from the floor makes their use more convenient when trunks, racks, and boxes line the walls. Adequate overhead light fixtures are necessary to ensure there are no dark corners. One central light in a 120-square-foot room may not be sufficient, as most enclosed ceiling light fixtures are limited to 60-watt bulbs for safety reasons.

Since leather goods are best kept at moderate temperatures and low humidity, the walls, ceiling, and floor of a tack room should be insulated. Even without a heater or air conditioner, insulation has the ability

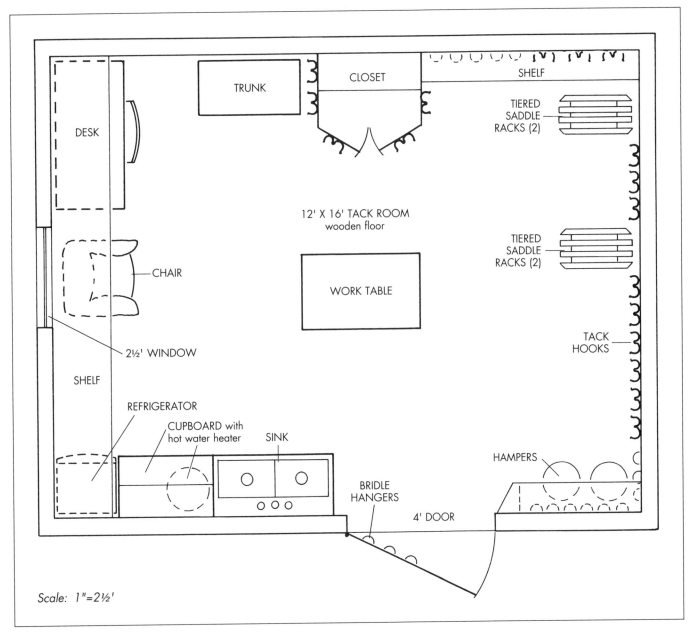

DESK

TRUNK

CLOSET

SHELF

TIERED
SADDLE
RACKS (2)

12' X 16' TACK ROOM
wooden floor

CHAIR

TIERED
SADDLE
RACKS (2)

WORK TABLE

2½' WINDOW

TACK
HOOKS

SHELF

REFRIGERATOR

CUPBOARD with
hot water heater

SINK

HAMPERS

BRIDLE
HANGERS

4' DOOR

Scale: 1"=2½'

to keep the indoor environment more constant. Some climates may require the use of a dehumidifier to keep mildew from forming on the leather during warm, wet weather. During winter months, a small space heater will keep the chill out of the air and prevent medicines from freezing. Be sure the heater is safe and does not present a fire hazard.

In areas with very cold winters, designing a freeze-proof cabinet may be better than heating the entire room. An insulated, heated cabinet requires far less electricity than is required to heat the entire tack room and keeps the assortment of veterinary supplies from piling up in the back porch of the house. The doors and walls of such a

cabinet should be insulated. A small, safe electric heater or large light bulb with a thermostat control set at about 40 degrees Fahrenheit will usually keep ointments, oils, aerosol products, and other items from being destroyed by freezing. Setting the thermostat at slightly higher temperatures will keep salves softer for more convenient application.

Although windows can provide a desirable air flow, sun rays shining through glass can be very destructive to leather. If windows are desired, those in direct sunlight should be outfitted with shades. Exterior windows, however, unless fortified with bars, decrease the security of the tack room. Most tack rooms are satisfactory without them.

Drawing #18
Tack room plan.

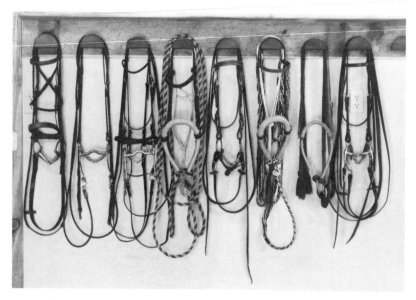

The tack room should provide ample space for hanging bridles and other equipment.

Tack hooks can be designed to hold many items. COURTESY OF CHERRY MOUNTAIN FORGE

Just because the room is being built in a barn, the carpentry should not be casual. Tight doorways and precise fit of wall to ceiling and floor will prevent infiltration of dirt, bugs, and rodents, all of which can be damaging to tack. The material for the floor should be durable, water resistant, and easy to sweep. Dirt floors are a constant source of dust, prevent tight wall-to-floor fit and defeat the purpose of insulating the rest of the room. Cement floors, although easy to keep clean, are very cold. An insulated wood floor is a very good choice.

Four-by-eight sheets of ¾-inch wafer wood (also called chip board, pressed wood, and aspenite) nailed to 2-inch by 6-inch floor joists on 16-inch centers work well. The wafer wood can be covered with a wide range of attractive, easy to clean, durable-resilient floorings such as vinyl tile and linoleum. Or you can just treat the wafer wood with a good floor paint, a commercial wood sealer, or a custom homemade mixture such as: 2 quarts boiled linseed oil plus 1 quart urethane plus ½ to 1 quart mineral spirits or turpentine. The variation in amount of turpentine used will alter the thickness of the mixture. Several light coats are better than one heavy coat, and the first coat or two should be of a thinner mixture. A boot scraper or mat, placed outside the door to remove mud and snow before entering the tack room, will help to preserve the floor.

Hanging gear. A large amount of wall space is required for hanging equipment, especially bridles and halter. Bridle holders with contoured head pieces that approximate the configuration and size of a horse's poll area help to keep bridles in good shape. A half-circle of wood about 4 inches in diameter and about 2½ inches thick works well. A ⅜-inch lip on the forward edge of the curved surface will keep the bridle from slipping off the bridle holder. Mount the wooden bridle holders 9 inches apart, center to center, to provide ample room for convenient use.

Other gear, such as running martingales, ropes, extra cinches, and nosebands, can be hung on hooks. Many types of commercial hooks are available, and custom hooks can be fashioned from, among other things, old horseshoes. Figure how many hooks you think you will need, then double the number! It is one of Murphy's laws that the pair of reins that you want is always at the bottom of a pile of entangled leather and nylon straps. Outfit your tack room and grooming area with plenty of hooks.

Saddle racks can be free standing or built-in. The former style is essential for the compulsive rearranger. Free-standing units are useful as saddle-cleaning stands and can be relocated near the heater in January or carried outside to a shady spot in June.

The touted convenience of fold-out built-in wall-style saddle racks has always managed to escape me. Does one hold on to the 40-pound saddle with one hand while stowing the rack along the wall with the other, only to be able to use the resulting space for the time while the saddle is in use on the horse? If you do choose wall-style racks, be sure to mount them with ample room in between — more than seems necessary. Multiple saddle racks, two or three high, make efficient use of space, but western saddles are difficult to lift on to a rack 6 feet above the floor.

No matter what type of saddle rack is chosen, be sure it satisfies the following requirements as outlined by your saddle! The rack should allow air circulation — the pipes or slats should be well spaced so that the sheepskin or panels and stirrup leathers can dry after hard use. The rack should be designed to approximate the contours of the horse's back. Although some horse owners may think their horse looks like a 55-gallon drum, using one for a saddle rack prevents the saddle from drying and spreads it out in an unnatural shape. Not many horse's backs are shaped like logs either. A saddle rack that supports the front jockeys and rear portion of the skirts without causing them to become misshapen is ideal.

Wet saddle blankets are often set to dry by turning them upside-down on top of the saddle. This might work well during hot, dry weather or if the blanket does not need to be used again soon. Faster drying time can be achieved by locating a blanket drying rack where it can take advantage of the sun and/or air currents.

Tack trunks are handy for storing items that are not used frequently or for seasonal items such as storm blankets. Trunks take up a lot of floor space, however, so it is best if they can double as seats or short-term work areas for simple repair jobs or small cleaning jobs. If you can afford it, plan to put a sink and small hot water heater in your tack room so that you can do your tack cleaning frequently and conveniently. Also, be sure to have a fire extinguisher in your tack room, and know how to use it.

A place for everything and everything in its place. Shelves, cupboards, and cubby holes will help to remind you to return an item to its spot. This way gear stays cleaner, lasts longer, and is there when you need it. Bins work well for small items such as bandages, protective boots, spurs, gloves etc.

It is convenient to use the tack room for veterinary, farrier, and training records. A corner of the room can serve as an office. Outfit a small desk with a good light, your

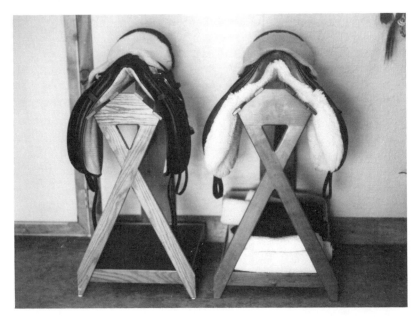

These English saddle racks allow the saddles to dry, while retaining them in the shape of a horse.

files, pencils and pens, a calendar, and a coffee pot!

The Tacking/Grooming Area

Just outside your main tack room door should be an open area specifically set aside for grooming, tacking, clipping, and shoeing. The floor should be a level, non-slip surface, and there should be a safe place to tie. Whether you use a post for tying or cross-ties is personal preference — whichever you choose should be well-designed and sturdy. If you use a tie rail or post, it should allow you to tie the horse at or above the level of the withers. If you choose a cross-tie, locate the rings at the height of the horse's head.

There should be plenty of conveniently located receptacles for clippers, vacuums, and your radio. And there should be a shelf or bins for your most commonly used grooming tools and supplies. The shelves should be located where the horse cannot knock down items with his mouth or tail, especially during clipping and shoeing.

Feed Room

Although the bulk of your hay should be stored in a separate building, you should

Nylon ties with a bull snap on one end and a panic snap on the other. These can be used with tie rings, cross ties, or in trailers. COURTESY OF VETLINE AND BMB

To keep the barn aisle clear, you may wish to locate a grooming and tacking area off to one side.

have a small feed room in the barn for a week's worth of feed. Be sure the room is separated from the stalls and has a horse-proof lock on it. Laminitis (founder), one of the most debilitating diseases of the horse's hoof, often begins with a grain overload from a horse getting into the feed room.

The feed room should also be rodent-proof, dry, and well ventilated. Concrete floors are fine. Grain should be stored in covered containers, such as large garbage cans. A 30-gallon garbage can will hold 100 pounds of corn; a 50-gallon garbage can will hold 100 pounds of rolled oats. Do you see once more why you should feed by weight and not volume? A table or shelf will serve as a place to store your buckets and measures and to weigh and mix up rations. You should have both a hay scale and a grain scale.

In addition you will need hay carriers, and hay nets if you use them, and wire cutters or a knife for opening bales. You will want to sweep the feed room out thoroughly every week before you bring in the new week's worth of hay. Along with your broom,

you may want to store your manure fork, shovel, rake, cart or wheelbarrow, manure basket, barn lime, and other stall-cleaning equipment. The feed room is also a good place for a small tool kit, a flashlight, and a fire extinguisher.

Water

Besides the waterers in the stalls and the sink in your tack room, you will want some additional hydrants and faucets in and around the barn for convenience. Pipes leading to the stable should be buried three feet or more depending on the depth of the frost line in your locale. One of the places you will want water, and warm water if you can afford it, is the wash rack and/or laundry area. This can simply be a pad of scored cement with a central drain and a safe place to secure the horse either by tying or using stocks. The washing machine can empty into the wash rack drain or can empty into a septic system if you were required to install one for a barn lavatory. If you use a wringer washer, you can simply attach a drain hose to it. If there is no septic system involved in your barn, the drains from your sink, wash rack, and washing machine can lead to a drywell, a large hole filled with gravel that is used to dissipate water into the soil. A drywell is much cheaper than a septic system if your building code will allow it.

Aisle-Ways

Aisles should be a minimum of 8 feet wide, but 10 to 12 feet is better. Any wider is a waste of space. In the aisle there should be safely designed and well-located hooks for halters and lead ropes, and storage racks for blankets. Be sure not to hang any items where a horse can reach them.

CHAPTER NINE

Other Buildings

DEPENDING ON THE SIZE AND SCALE OF YOUR operation, you may find the need for various buildings other than a barn for the proper functioning of your acreage. As you develop your property, you might wish to consider adding a farm equipment shelter, hay barn, loafing sheds, or other specialized buildings.

Hay Barn

Due primarily to fire hazard and dust continually falling into the stalls, the traditional hay loft over the stalls is no longer recommended. It is now recommended to separate the housing for horses from the storage of feed and bedding.

Location

The hay barn should be located separately from the stable yet close enough for convenience in transporting weekly supplies of hay, grain, and bedding to the feed room in the barn. To determine your hay storage needs, figure you will need between three and four tons per horse per year if you feed hay year-round and do not rely on pasture for supplemental feed. A ton of hay, 2,000

pounds, is usually comprised of about thirty 65-pound bales. A ton of hay requires approximately 200 cubic feet of storage space. For example, this would require a space 10 feet by 10 feet and 2 feet high; or 5 feet by 5 feet and 8 feet high; or 6 feet by 6 feet and 6 feet high.

The hay barn or open shed should be located on a well-drained site. Because of the labor involved and the cost of hay, you simply can not afford to have moisture get into your hay barn. You may have to incorporate ditches, slopes, or retainer walls to prevent water from collecting around or moving through your hay barn or shed. In addition, the roof must be leak-free and there must be adequate ventilation in the hay barn.

Curing Hay

Newly baled hay should be allowed to cure for several days in the field (or a week or two in an outside stack if necessary) before loading into the barn. This curing time allows the hay to "sweat" or release excess moisture. If too-wet hay is put into an enclosed building, the spontaneous combustion of the fer-

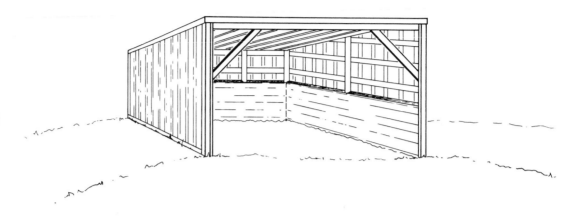

Drawing #19
12' x 20' steel walled three sided shed on well-drained site with eight foot back wall set to prevailing winds. Note solid half-walls with cap and metal edging.

mentation process may cause the hay to catch on fire. If when you walk into a hay barn you detect a damp or fruity aroma or a moldy or caramel smell, the barn probably contains spoiled or heating hay. A simple means of assessing the problem is sticking a hand into a bale. It should feel cool. If it is only slightly warm the problem is minimal and may not result in any spoilage or fire.

If, however, the bale is very warm or hot to the touch, take the temperature 2 feet into the stack. Push a pipe down into the stack and then drop a thermometer on a string into the pipe and let it hang there for about ten minutes. This allows the pipe and the air in it to attain the temperature of the stack. If the temperature registers over 160 degrees Fahrenheit, the hay should be removed from the barn and spread out to dry. It will no longer be suitable for horse hay, but could be used for mulch or cattle hay. See chapter 12 on pasture management for more details on assessing the quality of hay.

If grain and bedding are also stored in the hay barn, management practices should be employed to prevent spoilage or heating. In addition to keeping all grain in rodent-proof containers, it is a good idea to let the feline rodent patrol have access to the barn.

Outdoor Hay Storage

Although the best way to store a supply of hay is in a separate building, economics often dictate that the hay be stored outdoors, at least temporarily until the hay barn is completed. First select a level, well-drained site in a convenient location to distribute the hay for feeding. The stack can be placed to offer some degree of protection for animals from wind, but in cold-winter regions the protected side will usually end up with deep snow from drifting.

Rather than stacking hay on bare ground, place it on palettes, so the bottom layer will remain dry. Used palettes are often available free or very inexpensively from feed mills, lumber yards, or cement plants. In lieu of palettes, 2-by-4s or 4-by-4s can be set up side by side and the bales can be placed to span them.

Stack the bales tightly together, alternating the direction of the bales every two or three layers so that the stack is more stable. If you are planning to cover the hay, finish your stacking with a ridge of bales on the top rather than with a flat top. This will keep water and snow from accumulating on top and possibly leaking through.

In some geographical areas, an uncovered stack fares well with very little nutritional loss from sun or precipitation. In places with snowy, freeze-and-thaw winters, however, it is best to cover the stack—with a good tarp. It is better not to cover a stack than to cover it with a tarp that is full of holes. Letting water enter a covered stack is just asking for widespread molding on sunny days.

The best cover for a stack is a canvas tarp, as it is waterproof but allows for some air exchange to minimize condensation. Although new tarps are expensive, they will last a long time. You might be able to find a used truck or machinery tarp for sale at an auction. Black agricultural plastic also can be used except that it can tear or be punctured by the hay stems, it is difficult to tie down, and it can result in condensation under the plastic, causing hay spoilage. Blue polyethylene tarps may be initially inexpensive, but they have very low resistance to sunshine and often deteriorate in one season.

Coverings should be secured with twine or rope at all edges as well as over the top of the stack. Canvas tarps and some polyethylene tarps have sturdy grommets that are useful for tying down the edges of the cover. Improvise when covering with black plastic by placing a pebble or marble slightly in from the plastic's edge to create a lump to which to attach your twine. It is also advisable to run several ropes across the length and width of the stack to prevent billowing by the wind, which can loosen and tear a covering.

Cut up an old inner tube and use the rubber scraps under the rope or twine wherever it looks like it may cut into the covering when you tighten the rope. Holes in most coverings can be patched with a daub of silicone and a scrap of plastic or canvas. If you take a little extra time when you stack and cover your hay, it will have a better chance of retaining its quality throughout the year.

Shelters

If you have decided to keep your horses on pasture, they will need a place to stay dry and out of the wind. Wooded areas, slopes, and large rocks or ledges will provide good natural shelter. On flat land, one of the best ways to provide protection is with an in-and-out shelter or a three-sided shed. You may find, however, that in spite of your good planning and building, your horses choose to stand out in the snow and the rain! Such is the nature of horses. When it is wet and cold or windy they will seek shelter and greatly appreciate your efforts. During the summer, the horses may also use the shed as protection from the sun or flies. Because of its non-splash bedding, however, many horses tend to use the shed as a place for urination and defecation, and once habits are established, they are hard to break. If you are building a new shed, you may wish to close the horses out of it except during seasons in which they need it.

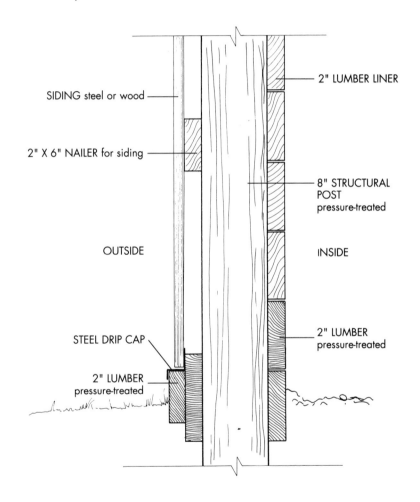

The opening of a three-sided shed should face toward the mildest winds, the back of the shed to the strongest winds. Sometimes Mother Nature makes this very difficult to determine. Usually a south or southeast opening provides winter sun and protection from the cold north winds. In any case, each locale and each spot on your land

Drawing #20
Wall construction of three-sided shed designed to prevent injury and preserve building foundation and walls.

may have different wind patterns that you must determine before selecting your site. Select one of the highest points in the pasture, unless it is the windiest spot. The shed should not be located so remotely that you rarely visit it. You must clean the shed regularly, check for damages, and tend to the horses daily.

You should figure about 140 square feet per horse, unless you have ponies or horses that are unusually compatible, in which case you can plan for considerably less space. The ceiling should have a 9-foot clearance on the low side and 11 to 12 feet at the entrance. If you attach an overhang of 4 to 6 feet, it will prevent some snow and rain from getting in as well as providing more shaded space.

Other Facilities

Depending on the scope of your horse program, you can build all sorts of specialized facilities. If you have more than two or three broodmares, you may wish to construct a separate barn or outdoor facilities for foaling and housing mares and foals. Your broodmare barn could have comfortable quarters for observers adjacent to the foaling stall. Or you could have a closed-circuit TV in the foaling stall, which you could monitor in the house.

If you plan to handle a large number of young horses, you may wish to have specially designed weanling and yearling barns and facilities. Special design features should include smaller spaces between fence rails and ample, safe room for the explosive outbursts of youngsters.

If you will be keeping a stallion, his facilities must be extra high and strong. His stall should be a minimum of 14 feet by 14 feet with 6-foot-high solid walls. His paddock should be located away and upwind from mares but close enough so he can see other horses. Stallions need very experienced, competent handling and extra attention. They can not be cooped up without exercise for very long before their energy gets potentially uncontrollable. Be sure to check zoning ordinances and local laws regarding the keeping of stallions.

Before you consider keeping a stallion, please realize that the only reason to keep a horse a stallion is, if he is a superior individual for breeding purposes. To determine this, have him evaluated by several non-biased professionals. If he is not worthy of being a stallion, have him gelded between the ages of six months and one year of age. (See *The Formative Years*).

If you will be standing a stallion for breeding, then you will need to develop some breeding facilities, which may include a breeding shed containing a tease rail, a mare preparation stall, a foal restraint stall, a phantom mare for collecting the stallion, and a laboratory to process the semen. The breeding farm is out of the realm of this book, but I have listed other books in the Appendix to help you make good decisions in this area.

CHAPTER TEN

Machinery and Equipment

....

ALTHOUGH YOU WILL PROBABLY NEVER ELIMINATE the need for a wheelbarrow and a fork, the size of your horse operation and its associated tasks may require some specialized machinery. Such equipment usually falls into one of two categories: a truck with a trailer and a tractor with its attachments.

Truck and Trailer

A truck and horse trailer is a must for an acreage horse-owner. First, in the case of emergency, you must be able to transport your horse to a veterinary clinic if necessary. For routine veterinary care, and especially if you live in a remote location, you can often avoid paying a "call charge" by transporting your horse to the veterinarian. If you plan to show or haul your horse to trail riding areas, you will also need a truck and trailer.

The Towing Vehicle

First you need to evaluate your present vehicle to see if it is suitable for pulling a horse trailer. What is its towing capacity?

Especially for interstate driving, the weight of a loaded towing vehicle should be at least 75 percent of the weight of the loaded trailer. Is the distance from the front axle to the back axle (wheelbase) of your towing vehicle at least 115 inches? The longer the wheelbase, the less the trailer will sway while in motion. Does your truck or car have a powerful enough engine to haul the extra weight of a loaded trailer? What you save in gas mileage with a smaller engine may result in greater repair costs, due to excess strain on the engine and transmission. Most often, a ½-ton or ¾-ton pickup truck is appropriate for pulling a two- or four-horse trailer, respectively. If you plan to pull a larger trailer, or if you plan to use your truck to carry heavy loads of hay or grain, you may need to consider purchasing a one-ton truck.

Does the towing vehicle have heavy duty springs and shocks (suspension) and good brakes and steering (power preferred)? Are the tires dependable and do you have a good spare, lug wrench, and jack? Are they easy to get to and do you know how to use them? Tires should be inflated with equal, optimum pressure.

Ideally, your truck or car should have a

special towing package, which includes a heavy duty radiator, a heavy duty transmission with special cooler and a low gear ratio, and heavy duty springs and shocks. Look in the owner's manual for the specifications on your vehicle or ask your dealer to help you with the answers to these questions.

Routine servicing of the engine, cooling systems, suspension, tires, wheel bearings, brakes, and other mechanical components not only prolongs the life of a vehicle, but also allows you or your mechanic to uncover problems before they become emergencies. Some common problems encountered with a hauling vehicle are overheated engines and transmissions, flat tires, and brake or hitch failures.

The Trailer

Trailer selection. There are many factors to consider when investing in a horse trailer. New trailers will cost from $2,500 for a bare-bones two-horse model to $10,000 and higher for a deluxe four-horse with dressing room and tack room. Many good new trailers can be purchased for $6,000, and used trailers can be found for substantially less. No matter whether your budget dictates new or used, there are basic decisions you must make.

There are three types of horse-hauling vehicles: the enclosed trailer, the stock trailer, and the horse van. Enclosed trailers usually come in two-horse and four-horse models and are the most common trailer seen on the road. The height inside trailers ranges from 72 inches to 90 inches, with many near 76; length of standing room ranges from 66 inches to 88 inches (depending on style) with the average somewhere around 70; the width of one stall is from 26 to 32 inches with most toward the low end. A sixteen-hand horse can fit into a standard trailer, providing he is level-headed about loading and unloading. Better suited for large horses are the 7-foot-high diagonal-load trailers which allow ample space for the tall and

long-bodied breeds. If you are planning to haul a very large horse, you may need to look into a custom trailer or van.

Stock trailers are usually the equivalent of a four-horse trailer in length and basic style, but the sides are slatted rather than enclosed and are designed to haul horses without partitions between them. This comes in handy for mares and foals and young horses. Horses have less of an enclosed feeling because they can see out of the stock trailer but because of the slatted sides, horses get very dusty, cold and wet.

Vans are horse stalls on trucks. Vans are more comfortable for the horse than conventional trailers and are the most expensive of the three types.

Trailer construction. The materials and workmanship dictate, to a large degree, the cost of a trailer. Materials commonly used include steel, aluminum, and fiberglass. A trailer with a frame and skin of steel is sturdy. Substituting an aluminum skin while retaining the steel frame will decrease weight and rusting. Fiberglass is often used for roofs and fenders as it is cool, lightweight, and easy to repair. Quality workmanship will be evident in the straightness of the frame, the fitting of seams, the finishing of edges, and the paint job.

Good-quality trailer suspension should be sturdy but not stiff. Whether you decide on leaf springs or rubber torsion suspension depends on the quality of each, but the latter gives a less stiff ride. Be sure the suspension is independent: that is, when one wheel hits a bump or a hole, it will absorb the shock independent of the rest of the trailer. Electric brakes are preferred and there should be adequate operating lights and clearance lights for nighttime driving.

Trailer design. You must decide how you wish your horses to ride in the trailer—facing forward, backward, sideways, or loose. Most trailers are designed for rear loading and place the horses side by side, facing forward. Other options are available if your horse is a difficult traveler. In many trailers, the center divider is removable to accommodate a large or difficult traveler or a mare and foal. If the center divider does not go all the way to the floor, your horse will likely travel more comfortably as he will be able to move his feet farther sideways to balance.

Horses load into a hauling vehicle by stepping up into the trailer or by walking up a ramp. The step-in style is less expensive and more common. The ramp style is usually safer and more expensive. The length of the ramp dictates the slope the horse must climb. A power-assisted ramp will be easier for you to close.

Trailer details. Are you looking for a bumper hitch trailer or a gooseneck style? A gooseneck trailer attaches to a "fifth wheel," a type of hitch that must be installed in the bed of your pickup truck. What type of tack room do you need? A small compartment for just a saddle and bridle or a larger one which can also be used as a dressing room? Consider the following options, realizing that for every one you add, the price will increase. How many vents, windows, and interior lights do you require? A minimum of one bus-style window on each side of the trailer is suggested. Does the center divider of the trailer need to be removable? Does the center divider go all the way to the floor? Do you want padding on the sides of the stalls, on the center divider, and at the chest? What type of flooring is available? Oak or pine are both fine as long as the quality of the wood is good—with no warping or knots. Pressure-treated wood may withstand manure and urine longer than non-treated wood. What type of mats come with the trailer? Removable rubber mats are preferred. What type of release bars are there at the chest or head, the tail, and at the center divider? Check to be sure all releases work easily. Some are very difficult to operate if the trailer is on less than 100 percent level ground.

Don't assume that just because a trailer looks good that it is safe. For example, a well known manufacturer overlooked one small detail in rear-door design which let the doors be easily lifted off their hinges. When a horse leaned against the rear door, it came off. The horse fell out of the trailer and was dragged behind it. Get competent help when you select a trailer.

Trailering can be very hard on a horse's muscles, bones, joints, ligaments, and tendons. Rough roads, long miles, inexperienced or inconsiderate drivers, inadequate trailer suspension, poor floor mats, and improperly maintained tires can all cause unnecessary wear and tear on your horse. To help assure that a horse arrives at its destination refreshed rather than fatigued, make the trailer as safe and comfortable as possible. Organize a maintenance plan for your horse trailer as you do for your towing vehicle.

Trailer maintenance. Store your trailer on level ground with the hitch jack adjusted so that the trailer's weight is balanced between the tongue and the tires. Park the trailer out of the weather to preserve the paint job, and on pavement whenever possible to protect against tire rot. If stored for several months, jack up each side and place blocks under the axles where the springs attach.

With the trailer hitched to its towing vehicle, assess its balance. The truck hitch should be at the right height so that the trailer floor is level. Most of a trailer's weight should be borne by its wheels. At a standstill on level ground, only about 10 percent of the weight of a bumper-hitch trailer should be transferred by the tongue to the towing vehicle. A trailer that is tongue-heavy will overstress the rear end of the towing vehicle and cause excess wear on the ball of the hitch. A trailer that is rear-heavy can essentially lift up the towing vehicle by its hitch and causes dangerous sway when the vehicle is in motion.

Evaluate the following major items at least once a year and repair or replace worn or broken parts.

Wheel bearings. The grease that lubricates wheel bearings picks up dirt and dries out. Have the wheel bearings cleaned and then repacked with grease annually. The seals will be replaced at the same time.

Brakes. The wheels should be removed so that accumulated dust and dirt can be removed from the brakes. Also, have the pads checked for wear and replaced if necessary. Several times each year, perform a brake check and adjustment. A gravel roadway makes a good area for this road test. It helps to have a knowledgeable observer to tell you whether a particular wheel is locking up or rolling free when the others are stopping properly. In either case, the brakes of that particular wheel will need to be adjusted.

Once all of the brakes of the trailer are stopping evenly, you can proceed to the next part of the brake test. Preferably on level ground, accelerate to about thirty miles per hour. Then without using your brake pedal, bring the rig to a stop by using the manual electric brake controller mounted on or under your dash board. If the trailer brakes grab, adjust the controller down. If the trailer can't stop the rig, adjust the controller up. Once the trailer brakes alone stop the rig properly, use your brake pedal, which activates both your towing vehicle brakes and your trailer brakes.

Tires. Rotate tires once a year to equalize wear. Check for bare patches, bulges, and other defects.

Suspension. At least once a year, grease springs and shackles. Check the bushings where the spring ends are pinned to the shackles. Check shock absorbers, if present on your trailer, and replace when necessary.

Floor. Check the floorboards at the beginning of the season for rotting, splintering, shrinking, or warping. Replace any boards that are remotely suspicious. Use clear (no knots) 2-inch-thick planks. You may wish to treat the floor with a preservative to combat the effects of manure and urine. Use resil-ient mats with "life" to help absorb road vibrations and shock. Replace mats when they have become excessively chewed up by shod hooves or compressed from long use.

Sideboards. The bottom two to three feet of the side walls of your trailer take a lot of abuse from the inexperienced or scrambling horse. If the walls are metal, check for rust. You may wish to install ¾-inch plywood kickboards over the metal walls for added protection for both the horse and your trailer.

General. Be sure that the hitch, safety chains, chest bars, tail bars, dividers, doors, and windows work properly. Each time you use your trailer:

● Sweep out the stalls of the trailer and remove the mats so the wooden floor can dry out. The mats can be hung over the stall divider.

● Clean mangers of old hay and grain to prevent mold (danger to the horse) and rust.

● Check the tire pressure. (It should usually be around 32 psi, but check for the specification on the sidewall of your tires.)

● Check to see that wheel lug nuts are tight.

● Check tires for irregular tread wear, bulges, defects, or weather checking.

● Check spare tire and jack and know how to use them.

● Oil any hinges, latches, or other moving parts that do not function freely.

● Check running lights, turn signals, brake lights, emergency flashers, and brakes.

● Check the inside of the trailer for such things as hornet or mouse nests.

● Add fresh bedding (sawdust) if desired.

● Wash trailer as needed and wax twice per year.

● Check the hitch and safety chains.

● Put trailer registration in the towing vehicle.

Trailering. If you are hauling only one horse, load it into the left side of the trailer, so that the horse is riding up on the crown of the road rather than on the shoulder edge. The weight and movement of a single horse in the right stall can cause the trailer to be pulled off the road onto the shoulder.

Traveling. Always allow yourself more traveling time than seems necessary. Know your route ahead of time. Give your truck and trailer another once-over to be sure that all systems are go: the hitch, lights, mirror adjustment, etc. Be sure you have a spare tire, lug wrench, and a jack to fit the trailer, and the trailer registration.

The horse is usually secured by a trailer tie, a 12-to-18-inch rope with a panic snap on one or both ends, which is secured to a tie ring in the manger of the trailer. A panic snap is designed so that it can be released even if the rope is taut. A normal snap would be impossible to unfasten in such a situation. To prevent the horse from stepping on the lead rope while in transit, be sure to remove it after the horse is loaded and attached to the trailer tie.

Once your horse is secured in the trailer, make a thorough last-minute check to see that all tack and trailer doors are latched, that the hitch is fastened properly, and that the lights are working. The hitch jack must be raised to its highest position so that it will not contact the road. Be sure that all wheel blocks are removed and put in the rig before you pull away. Once you are out of the driveway and on straight, level ground, recheck all of the mirrors for proper alignment.

On the road. Drive safely. Be considerate of your horse passengers. Accelerate and brake gradually. Anticipate stop signs and corners, giving yourself ample distance for slowing or stopping. Use your turn signals. Take curves at a moderate speed. Don't make any quick lane switches. Keep your senses alert to unusual sounds, smells, and vehicle motions.

It is possible that some time during one of your trips you may feel your trailer swaying from side to side. This can be caused by low tires, moving horses, wind gusts, road irregularities, or turbulence caused by passing vehicles. If you slam on your brakes, as is often the instinct, you will only worsen the situation. Instead, while continuing to steer straight ahead, take your foot off the accelerator and gently apply the manual trailer brake mounted on your dash board. This will allow you to regain control of the vehicle. It is a good idea for you to practice this exercise ahead of time with an empty trailer until you feel you have overridden your reflex to put your foot on the brake to counteract sway.

Shortly after you leave home, you should stop for a trailer check. Then plan to stop routinely every fifty miles or so. Give yourself a stretch while you walk around the rig. Check the tires and the hitch and take a look at the horses. Often during such a stop, you can detect a problem early, such as a soft tire or a hot brake, before it becomes a disaster.

In case of emergency. It is a good idea to routinely carry an emergency kit in your towing vehicle. It should contain flares, jumper cables, a dependable light, and some tools. Use the recommendations in your state's driver's manual for indicating a disabled vehicle.

Tractor and Attachments

If the activities on your acreage justify investment in a tractor and attachments, take plenty of time to research the market so that you buy what you really need. Determine if your needs would be better and more economically satisfied by hiring a local farmer to do the work for you. If you decide that your needs warrant the purchase of a tractor, give careful consideration to each and every attachment you are considering buying. I recommend that you purchase your equipment through a reputable dealer.

Keep in mind that operating farm equipment can be dangerous. A tractor is not a toy for children to play with. Be sure that every person who will be operating your farm equipment is well informed about safety practices and agrees to follow certain rules.

Ten foot disc with hydraulic lift and wheels.

In 1984, the Occupational Safety and Health Association (OSHA) instituted safety regulations for tractors such as roll bars and seat belts. The majority of small farm tractors, however, because of their age, are not outfitted with such safety gear. It is up to you to operate and maintain your machinery in a responsible manner.

Generally you can buy farm equipment through a dealer, by private treaty, and at auction. I discourage you from considering the last two options unless you already have experience with machinery and equipment. By working with a reputable dealer, you have a much better chance of ending up with what you really need. Use the services of a professional — you get what you pay for. Most dealers offer warranties on new equipment, some offer a limited warranty on reconditioned equipment, most offer repair service, and many, if you become a regular customer of theirs, will provide advice and answers after purchase.

If you are going to buy privately or at auction, get a professional opinion on the item you are considering purchasing. Is it mechanically functional? If not, how major are the required repairs? Are parts readily available? What is the item's value? As with any professional service, be prepared to pay for an appraisal. A dealer may be happy to provide a professional opinion if you are a regular customer, but cannot afford the time required to give free advice to everyone at the auction!

Tractor

Some of the most common mistakes made in purchasing a tractor include being uninformed, using price alone as a buying criteria, buying too small, and buying a fix-up.

Certain single features on a tractor can make the tractor totally unsuitable for your use. If you are not aware of these things, you may end up with a tractor that seems 100 percent right for you but is actually about ten percent wrong—and often that ten percent will make the tractor unusable. For example, tractors with narrow front ends (a single front tire or front tires 6 to 8 inches apart) are highly maneuverable but very unstable. They have a reputation for tipping over, one of the main causes of injuries and deaths on the farm. A narrow front-end tractor is a risk in any situation, and should be avoided, but is absolutely unsuitable for hilly or mountainous use.

If you approach the matter of buying a tractor solely with price in mind, you may pay a low initial price but end up replacing or repairing many costly items on the tractor. You should not consider purchasing a fix-up project unless you are very experienced with tractor repair and parts availability.

Within reason, it is better to buy a more powerful tractor than your current needs require because you will almost always find more demanding work to perform with your machine. It is a costly mistake to buy a small tractor and then try to accomplish work more suitable for a big tractor. The mechanical workings of the small tractor simply will not hold up and you will be faced with extensive repair bills. No single tractor will do all jobs for you perfectly, but try to choose the type of tractor that is best suited for your needs.

Tractors can be loosely grouped into three categories as dictated by size, weight, engine size (horsepower), and suitability for purpose:

Lawn and garden tractor. (Approximately 16 horsepower). Only consider a light-duty tractor like this if you are planning to buy two tractors. A lawn and garden tractor is handy for driving through barn alleyways and pens or for pulling a small manure or feed cart. It would be totally unsuitable, however, for any field work, arena work, or large-scale feeding or manure handling. And these are expensive tractors for horsemen — by the time you buy a tractor, a cart, and other attachments, you have reached the same price range as in the following category. Essentially, you end up with half the tractor for the same price.

Acreage tractor. (Approximately 25 to 27 horsepower). These are convenient, easy-to-operate tractors. Since they are not very tall, they are pretty easy to mount. The hitch is low to the ground, making attachment of implements convenient. Generally, these are good tractors for teenagers to learn on.

Mid-sized tractors in this category are the equivalent of the 8N Ford, which was known as the estate tractor when it was manufactured (1939 to 1952). As a thumb rule, you can probably find a fairly decent used tractor in this category in running order for approximately $2,000. There weren't a lot of mid-sized tractors manufactured through the late fifties up to the seventies, so the vast majority of used tractors in the acreage size are at least twenty-five years old. Now the new mid-size tractors, which are referred to as compact tractors, are Japanese-made,

many of them four-wheel-drive. They go for $10,000 new and for $5,000 to $6,000 used (manufactured in the late seventies and early eighties).

Mid-sized tractors are good for all-around small acreage chores, but they are limited to the size of the attachments that can be used with them. They can pull about an 8-foot pull-type disc or a 6-foot three-point disc. They work well with a small manure spreader, especially the friction-drive type. (See manure spreaders below.) With a front-end loader or a 6-foot blade on the back, a mid-sized tractor works well for cleaning out pens and runs.

Farm tractor. (Approximately 40 to 70 horsepower). These are taller, more powerful tractors, able to operate more heavy-duty equipment. If you have a large arena and want to use a disc with two 8-foot sections, you will want to check out this size tractor. If you need to handle large amounts of manure and you use a heavy-duty spreader, you will need a tractor of this size. Depending on economics at the time you are ready to purchase, you may well find a larger tractor for the same money that you would pay for a mid-sized tractor. All other things being equal, if you need the extra power, buy the larger tractor.

Selecting a Tractor

Make. There are some types of tractors to avoid because they are so rare that it is difficult, if not impossible, to find parts for them. Any narrow front-end tractor should be avoided. Dry-land tractors, those without a three-point hitch, are very impractical. Buying an old Minneapolis-Moline for $200 may seem like a great deal, but when you realize that the draw bar on the back limits you to using a pull-type disc, it might not seem so wonderful. And when you try to find parts to repair it, you will find them very difficult to locate. Generally, most models of Ford, Massey Ferguson, John Deere,

420 John Deere with tricycle front. COURTESY OF RON'S EQUIPMENT COMPANY

A Farmall with adjustable front end, adjusted wide. COURTESY OF RON'S EQUIPMENT COMPANY

International Harvester, and Allis Chalmers still have parts available and would likely make good choices.

Condition of engine and drive train. How well a tractor starts is not critical, as a tune-up will often make a great difference. And basing your entire opinion on the level and cleanliness of the oil on the dipstick is a mistake also. Instead see how well the tractor runs and performs work you intend to use it for.

If the engine makes unusual noises, lugs, smokes, or leaks large amounts of oil, beware. As you are test driving the tractor, check the clutch. A worn-out clutch is usually caused by operator error—inexperienced drivers riding the clutch. There are two ways to check the condition of the clutch. First, with the tractor running, push in the clutch and put the tractor in gear. If it grinds going into gear, this usually means the clutch is on its way out. Next, put the tractor in low gear, let out the clutch and put your foot on the brakes. The tractor's engine should lug down and die. If, instead, the clutch slips, and the tractor continues running, it means the clutch is going out or gone.

Oil pressure. Check the oil gauge and be sure the pressure registers in the optimum range. If the gauge does not work, have a mechanic test the engine's oil pressure for you. Beware of tractors that smoke a lot as this indicates internal engine wear, often requiring big bucks to fix. There should be no knocking or rattling noises in the engine or the transmission.

Appearance. Don't be overly influenced by the appearance, as fresh paint can often be used in an attempt to cover up significant problems. Ninety percent of all tractors are stored outdoors. This means that while many tractors are functionally sound, they may look weatherbeaten, with oxidized paint and tattered seat cover. Some of these items may be minor, but one to look out for is weather-checked tires. (See "Tires" below.)

Power take off. The PTO is a revolving shaft on the back of the tractor that accepts attachment by a variety of equipment. It provides power from the tractor's engine to run such things as a manure spreader, a post-hole digger, a cycle mower, a brush hog and other implements. The PTO turns at 540 revolutions per minute and has been the cause of many farm accidents. An operator's fingers or clothing can get caught in the knuckled joint of the PTO resulting in serious injuries and death. Modern tractors and equipment have shields or guards around dangerous parts of farm machinery like the PTO, but such safety requirements were not in existence when the older models were manufactured. Therefore, if you use a PTO, be sure to keep children away from the equipment and use caution when working around it yourself. If you have an unshielded PTO, have a blacksmith or welder fabricate guards for your machinery.

Three-point hitch. A three-point hitch is the linkage on the back of the tractor that hydraulically raises and lowers the attached equipment. Most modern "acreage" tractors have what is called a Category 1 three-point hitch, which makes for more convenient attaching and operating of various pieces of farm equipment such as discs, plows, post-hole diggers, and so on.

Tires. Some weather checking is normal for tractors as they are stored outside, exposed to the deteriorating effects of mud and water, and, on used tractors, the tires are often quite old. If the weather checking has become so severe that it results in deep fissures, beware, as those tires are likely to come apart at any time. Front tractor tires can be purchased in the price range of car tires, but rear tractor tires will be somewhere around $150 to $200 per tire. When considering a tractor's price, therefore, don't overlook the condition of the tires as it may greatly affect whether you are getting a good deal or not.

Within reason, tread is not as important on tractor tires as it is on your car or truck. However, try to find lugs with at least 50 percent of their tread depth left.

Year. Don't be overly concerned about the year of a tractor. Because we have become conditioned to want a new car or truck every

Reconditioned B Allis Chalmers with belly blade and without three-point hitch. COURTESY OF RON'S EQUIPMENT COMPANY

so many years, we often look at other vehicles that way too, but with tractors, that is false economy. If it runs and it is paid for, the extra age of a tractor just adds to its character! And the quality of workmanship is really fine on most older tractors. There are a lot of very usable, good tractors from the 1950s and '60s. Just be sure that parts are available for repair.

Good equipment retains its value and, in some cases, actually appreciates. In 1948, a brand new 8N Ford was sold right off the showroom floor for $900. That same tractor is worth $2,000 today.

Warranty. New tractors usually have a complete one-year warranty for parts and labor and a two-year warranty on the engine and the drive train (clutch, transmission, rear end).

Used tractors are usually sold "as is," that is, "buyer beware." Some dealers, however, will offer reconditioned tractors for sale with a ninety-day parts and labor warranty. Although you will pay more for a reconditioned tractor than an "as is" tractor, you will have the peace of mind and financial assurance that if anything goes wrong, it will be fixed. Most things that are going to require fixing seem to turn up within the first few weeks of operation. An "as is" 8N Ford might cost $1,500 while the same one reconditioned, from a dealer, with a warranty, might cost between $2,000 and $2,200. Dealers have to add at least a 25 to 30 per-

The Power Take Off (PTO), which turns at 540 revolutions per minute (RPM). COURTESY OF RON'S EQUIPMENT COMPANY

cent margin to the price for their risk.

Preventive maintenance for your tractor.
For every dollar spent in maintenance, ten dollars are saved in repairs. Completely service your tractor every fifty hours of real working time. This can be monitored by the engine hour gauge or by using a log book. When you service your tractor, change the oil, the oil filter, the fuel filter, and the air cleaner (oil-bath filter), and grease all zerks (fittings). Always keep an eye on hydraulic fluid levels.

If you are in a temperate climate, every fall, be sure the tractor is winterized for well below the minimum temperature your area usually experiences. This maintenance item is not to be taken lightly—the damage that occurs when an engine block freezes and breaks is very costly to repair.

Attachments

Avoid package deals unless they contain just what you need. Don't buy an attachment just because the person selling it says it comes with the tractor. If you don't need a particular implement, ask what the price of the tractor would be without it. Choose only those items that are in very good working condition. When in doubt, get professional advice.

Disc. A disc is handy for lightly working the soil in an arena or for aerating a very compacted pasture. Pull-type discs are common and inexpensive because they are very difficult to transport. They cannot be raised and lowered like hydraulic or three-point discs. To be moved, pull-type discs either have to be lifted with a loader onto a trailer or dragged behind the tractor, discing everything along the way including driveways, road surfaces, and grassy areas! In addition, if you use a pull-type disc to work an arena, you will not be able to back deep into the corners and will end up with an oval area of worked soil.

Hydraulic discs are either raised by the tractor's three-point or by a separate hydraulic pump. Smaller hydraulic discs (6-foot) are operated off the three-point. They are raised in the air to be carried from point A to point B. Larger, heavier discs have tires that are lowered to accept the weight of the disc for transport. Hydraulically-raised discs prevent damage to areas that you do not want worked. They also allow you to position the disc in tight spots. Using reverse gear, you can raise the disc, back it into the corners of your arena, then set it down and work the earth right up to the fence line.

A pull-type disc has one point of attachment to the tractor's draw bar so it can sway and bounce somewhat. Three-point and hydraulic pump discs are held more rigidly in place than are pull-type discs, and consequently they pull harder and require a tractor with more horsepower. An average 6-foot three-point disc will cost about three times as much as a pull-type disc. Both, however, do an equally good job of discing.

Old discs with box bearings should be avoided because box bearings are very difficult to find today. It is best to stick with discs that have sealed bearings.

Harrow. Harrows are useful for smoothing an arena after discing, breaking up and spreading manure in pastures, and aerating compacted soil.

There are basically three types of harrows, or drags as they are sometimes called: the chain (or English), the spike tooth, and the spring tooth. The English harrow is made of heavy bars that criss-cross each other in a diamond-shaped configuration and have protrusions called teeth on the bottom side. They are very heavy and expensive but do a wonderful job of smoothing rough spots. They are good for leveling manure in a pasture as well as for aerating the soil without ripping it up. Homemade drags, simulating the English style, have been made with chain-link fence, but the lack of teeth and their light weight make them bounce on top of the soil and result in little smoothing and leveling. Like pull-type discs, English harrows are difficult to load, and when you move them by dragging them behind the tractor, they work everything in your path.

The teeth of the spike tooth harrow are adjustable for work or transport. You can use them in a full upright position to rip up pastures in the spring or you can set the teeth in a flat position to move the harrow or to use it as a leveler and smoother. After long or hard use, the teeth will become rounded and/or short but can be replaced. Some spike teeth bolt on to the harrow and some clamp on.

The spring tooth harrow is a cross between a ripper, a mild plow, and a harrow. Its configuration allows it to rough up the earth but it does not do a very good job of smoothing the soil.

Blade for the tractor. The typical horse farm blade is a 6-foot rear-mounted three-point blade. The blade can be mounted at an angle or swiveled around so that it can be used either to push or to pull. A blade is handy for scraping pens, leveling driveways, and moving light snows.

Loader for the tractor. Double-action hydraulic buckets are much more desirable than trip buckets. Pipe frame trip-bucket loaders, common in the 8N Ford era, raise and lower hydraulically, but operation of the bucket itself is an all-or-none situation. Because of this lack of finer control, trip buckets are notorious for either skimming over the top of what you are trying to load, or digging in too deep and gouging the earth underneath.

Double-action hydraulic buckets allow you to adjust the position of your bucket to a much greater degree. In addition, you can sprinkle the material you are unloading rather than just dump it in one pile as is the case with a trip bucket.

With any loader, look for signs of stress: bowed arms, welded repair spots, worn-out pins and bushings, leaking hydraulic cylinders.

Manure spreader. Manure spreaders are wagons with a mechanical apparatus designed to distribute manure as the tractor is driven through a pasture or field. The smaller, older spreaders are friction-driven; the bigger, newer spreaders are made to be powered by a PTO.

Chain drag (English harrow).

Friction-drive spreaders are ground driven, the power for the mechanics of the spreader being generated by the tires rolling on the ground. There are two levers, one to control the speed of the apron chain, which moves the load toward the rear of the spreader, and the other to control the beater bar at the back of the spreader, which flings the manure into the air. This type of spreader can be operated behind a pick-up or a team of horses, as it is a self-unloader.

Spreaders powered by a PTO are usually bigger, heavy duty spreaders suitable for a commercial farm. They can be more difficult to hook up.

Mower. You may wish to invest in a sickle bar mower in order to mow weeds. If your pasture will require routine clipping, a rotary mower will be more appropriate. If you have brush or heavy weed to remove, a brush hog would be suitable.

STORAGE SPACE REQUIREMENTS FOR MACHINERY

Equipment	Square feet
TRACTOR (25-40 horsepower)	70-80
PLOW	25
DISC (14 feet)	140
HARROW (14 feet)	100
MOWER	25
MANURE SPREADER	130
BLADE	25
POST-HOLE DIGGER	20
CART	20
TRAILER	160
BALER	100

Post-hole digger. Used ones are hard to find in good shape because either the augers are worn or the gear boxes are shot. You can't rent the type of post-hole digger that is used with a tractor, although two-man gas-powered ones are usually available for rent. If you are just beginning to put in your facilities and see lots of fenceposts in your future, it would pay you to purchase a post-hole digger for your tractor ($350 to $500). Otherwise, it would be more economically sound to hire someone to dig the holes for you ($1 to $5 per hole). Or you can dig the holes by hand, as you will still need a hand post-hole digger to clean out the bottom of the drilled holes.

Cart. You may find a two-wheeled cart handy to haul feed during chore time. This is especially suitable with a garden tractor.

Flat-bed trailer. The right type of utility trailer would be suitable to use with both your tractor and your pick-up. A common size is 16 feet long and 6½ feet wide, with tandem axles. The weight limit on such a trailer is 7,000 pounds, so if you wish to use it to haul hay, you can transport three to four tons at a time.

Hay baler. This implement is not warranted unless you are putting up hay on a minimum of forty acres. Figure on spending approximately $1,500 for a baler and the going price for a mower, rake, and field wagon. Add to this the cost of gas, baling twine, and your time. Then compare your total to the rates charged by a custom baler. He will quote you something like $1.50 per bale or offer to bale your hay on shares. Share-cropping usually involves you supplying the field of hay, the custom farmer supplying the equipment, supplies, and labor, and the two of you splitting the crop.

Balers have a deserved reputation for mechanical quirks. They are amazing and intricate machines with many things that can go wrong. If a piece of equipment is going to break down on your farm, it will be the baler. They are difficult to understand and trouble-shoot, expensive to work on, and if you don't know anything about farm equipment, they will frustrate you! So, unless you can really justify baling your own hay economically, stay away from balers.

Pasture sweeper. A recent innovation for horse farms where many horses graze in a limited space is the pasture sweeper, pulled behind a tractor. With a 5-foot sweeping width, the machine can clean up to 90 percent of the manure in a ten-acre flat pasture in four hours. This increases grazing area and decreases parasite infestation.

CHAPTER ELEVEN

Fencing Your Training and Turn-Out Facilities

WHEN IT COMES TIME TO CHOOSE FENCING, GET ready to do some research. In the last ten years, a great many new types of fencing for horses have appeared, and more are probably being invented as I write this. No single type of fence will be suitable for all of your plans, furthermore, and you may find it perfectly logical to have five or more types of fencing on your horse acreage. The ideal fencing will be different for particular needs, such as pens, paddocks, runs, pastures, round pen, arena, and so on.

Good fencing serves many purposes: Fences keep horses separated and in a particular place, away from the residence, lawns, crops, vehicles, buildings, and roads. Fences maintain boundaries and property lines, and thus promote good relationships between neighbors. Fences decrease liability because they lessen the chance of a horse doing damage to others' property, decrease the chance of a horse getting on the road and causing an accident, and keep people (especially children) and animals (especially dogs and other horses) off the property. Good fencing is designed to keep horses from getting hurt, whether the horses are turned out or being trained. And, finally, attrac-

tive fencing can set off an acreage and add to the value of the property.

As you choose fencing materials, your first consideration should be that the risk of injury is greater with horses than with other livestock. Since a horse's main purpose is movement, leg injuries, frequently associated with fence accidents, can put a horse temporarily or permanently out of service. Safe fences for horses are well made of materials that are sturdy, low maintenance, highly visible, attractive, and affordable. *Barbed wire is not a suitable horse fence.*

Planning

When laying out fence lines, avoid acute angles that can cause a horse to become cornered by other members of the herd, even if only in play. When running, whether from fright or exuberance, horses will try to go through or over fences. Four and a half feet is the absolute minimum fence height that will discourage horses from jumping. Five to six feet is better, especially for stallions, the larger breeds, or those specifically bred and trained for jumping.

To make a fence plan for your acreage,

FENCING YOUR TRAINING AND TURN-OUT FACILITIES **91**

Good fencing adds value to your acreage.

A protective cap for metal T-posts that prevents injury to horses and also serves as an electric fence insulator.

STRENGTH OF 7-FOOT POSTS SET 3 FEET DEEP

Size and Type of Post	Breaking Force Required In Lbs.*
WOOD 3" diameter	400-700
WOOD 4" diameter	800-1,500
WOOD 5" diameter	1,400-2,700
CONCRETE 4" diameter	295-310
METAL 1" T POST	152-160 (results in 1-inch permanent bend, not a break)
*Force exerted at the top.	

draw a scale map of your land on graph paper. Design a complete perimeter fence with the entrance gates always closed so escaped horses are kept on the property. Locate on the map all permanent structures, including such things as trees, water, and buildings. Draw current and proposed traffic patterns. Locate gates, and then draw in fence lines. Using rocks or boards as markers, translate the proposed plan from paper to the land itself. Make adjustments. When you have a layout that will work, mark your post holes. Using a taut string, mark the fence line. Finally, calculate the number of feet and amount of materials necessary: corner posts, brace material, line posts, gates, fencing material (such as planks, rails, rolls of wire), and miscellaneous supplies such as bracing wire, insulators, and electric wire.

Posts

Although the type of posts that you use will depend on the style of fencing you choose, some information regarding posts relates to nearly all fencing systems. When setting posts, be very certain that the line for the fence is absolutely straight; or it will cause you repeated problems both in the attachment of the fencing and with future maintenance. Use a carpenter's building line to set up the line for your posts.

Most posts are 8 to 10 feet long and are set 2 to 3 feet in the ground. Wooden fenceposts should have their tops sawed off at a slant to allow water to run off, but care should be taken not to make the angle so severe that it results in a dangerous point. A horseshoeing rasp can be used to dress the edges of the cut post tops.

Metal posts are usually set by driving

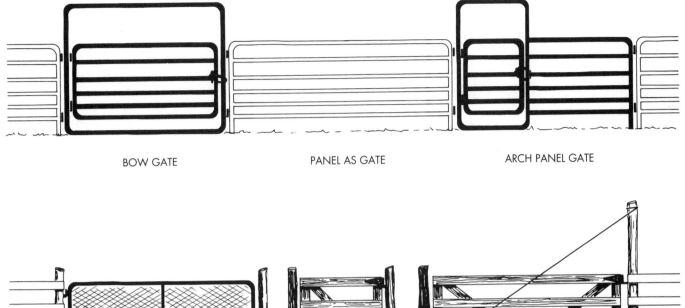

BOW GATE   PANEL AS GATE   ARCH PANEL GATE

STEEL FRAME WIRE MESH GATE   WOODEN MAN GATE   WOODEN EQUIPMENT GATE
with optional cable support

them into the ground with a manual or tractor-driven post driver. Other posts require that a hole be dug by hand, with an auger on a tractor or by a gas-driven post-hole digger. Posts that are set in holes are often secured at the bottom of the hole with some concrete or rocks before the earth is tamped around the post.

Posts are usually 6 to 12 feet apart depending on the fencing material. Some material comes in continuous lengths so the distance between posts is not as critical. But if you are using lumber or other materials that come in pre-cut lengths, the distance between your posts must be accurate. It is better to make the distance between posts a few inches less than the lumber lengths, so that you can trim off any split ends of boards.

Most fences have either a continuous sheet of fencing (such as woven wire) or from three to five rails, pipes, or wires, the bottom element being at least 1 foot from the ground. Continuous fencing can be stretched with a fence stretcher, a come-along, a truck, or a tractor. Fencing material should always be put on the inside of the pen, that is, toward the horse, to prevent them from contacting the posts and from pushing the fencing off.

Gates

Gates should be free-swinging, light to handle, and roomy enough for the intended purpose. Man/horse gates should be 4 feet wide; machinery gates between 12 and 16 feet wide. Walk-throughs are small spaces in the fence where a person can slip through sideways. These can be handy but may also be a hazard for a horse. Gates used with rigid metal corral panels are usually either a panel, a bow gate, or a gate in an arch frame.

Materials for gates vary as much as for the fences themselves. Wooden gates are

*Drawing #21
Gates.*

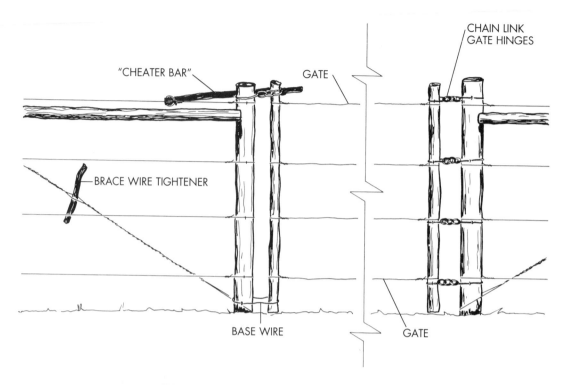

CHAIN LINK GATE HINGES

"CHEATER BAR"

GATE

BRACE WIRE TIGHTENER

BASE WIRE

GATE

GATE

Drawing #22
Gate and fencing tips:

Use a stick to tighten and anchor brace wire.

Use a "cheater bar" to make closing long wire gates easier.

Use chain links as hinges for long wire gates.

Staple base wire (for gate foot) to fence post to hold loop in position.

TOP AND BOTTOM:
Various means of securing gates.
BOTTOM PHOTO COURTESY OF
COLORADO KIWI LATCH COMPANY

often very heavy unless properly engineered, and it is difficult to keep a heavy wooden machinery gate from sagging. Heavy gates should always have a support block at both the open and closed positions to prevent sagging and to extend the life of the hinges. Some of the tubular metal gates are light, sturdy, and fairly safe. Flat metal gates with sharp edges should not be used on horse facilities.

Types and Characteristics of Fencing Materials

Wood

Wood fences are traditional and have a certain aesthetic appeal for a horse farm. If they are well installed and maintained, wooden fences increase the property value. Wooden fences are strong if put up correctly, with joints of board fences meeting tightly on the inside of the pen. Visibility and safety are good, except when broken or splintered boards and exposed nails cause accidents.

Drawing #23
Three types of walk-throughs for people.

Wood should be kiln-dried and treated with a nontoxic wood preservative. Wood is used as posts, boards, poles, and as the rails in buck fence. Posts should be a minimum of 4 inches in diameter up to 8 inches in areas requiring great strength. Wooden posts should be 8 to 10 feet long, set 3 to 4 feet in the ground. Board fences are often made of three or four 8-foot-long 2-inch by 8-inch pine boards as pine tends not to splinter. Pole fences, post and rail, and buck fences utilize 8- to 12-foot poles of varying diameters, peeled or unpeeled, treated or untreated with three or four rails. Sometimes pole and split rail fences are made of cedar, which weathers to a gray color and is resistant to decay.

Wooden fences do require regular maintenance. They should be checked yearly to assess the need for paint or preservative. Broken, splintered, chewed, or rotten boards need to be replaced. Creeping nails need to be driven in — expansion and contraction of the wood causes the nails to creep out, sometimes protruding an inch or more before you notice. Rails can fall or pop out of the mortises in posts and the post can split where the mortise has weakened it. Wooden fences must be protected from the chewing habits of horses with special preservatives, with metal strips on the edges, or with the addition of an electric fence.

Wood preservatives. Wood should be preserved to protect it from dampness, weathering, termites, fungus, and chewing. Look for specific marks on treated lumber or ask the dealer to determine how it was treated. You can also treat wood after it has been installed, usually with one of the oil-borne preservatives such as carbolineum. An oil-borne preservative is needed for wood that will be in contact with the ground, where conditions are ideal for rotting, termite damage, and fungus, and for wood that will be subjected to extreme conditions above ground.

Creosote. This distillate of coal tar contains some strong acids and will burn the skin, so extreme care must be taken on application. A dark brown substance with a strong smell, it should never be used indoors, although it has some outdoor applications. Creosoted surfaces cannot be painted until the creosote has weathered a year or more. Some leaching occurs but, in general, the product has a very long life expectancy. Due to the dangerous nature of the product, the government has banned the use of creosote without a license.

Penta (pentachlorophenol). This crystalline organic compound was developed specifically for the wood preservation industry.

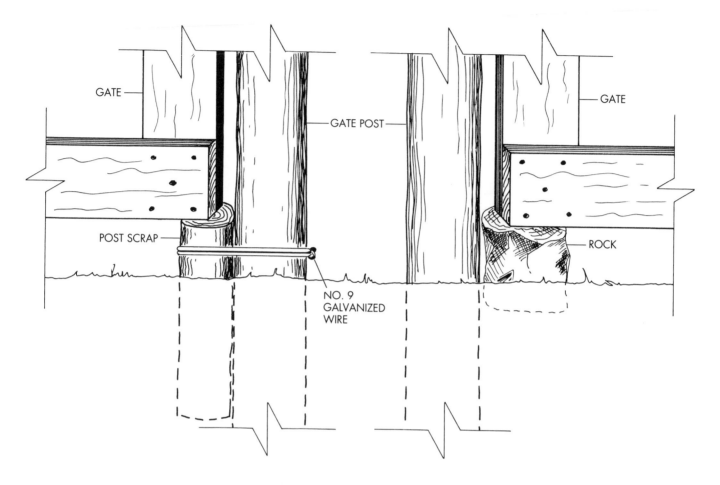

GATE

GATE POST

GATE

POST SCRAP

ROCK

NO. 9
GALVANIZED
WIRE

Drawing #24
Install supports to prevent gate-sag. Therefore, hinges and hinge post carry the load of the gate only when the gate is being operated. Install a support at the gate's open position, too.

Penta is normally dissolved in petroleum oil and forced into the wood with pressure or applied topically. Another commercially used wood preservative is *copper napthenate*.

Carbolineum. This blend of oils does not contain penta, arsenicals, or creosote. It is designed to be safely used on trees of a 2-inch diameter or more, as well as on fences and interiors. Carbolineum is available in clear or dark brown.

Buck Fence

A buck fence is ideal for rocky or very irregular terrain where post holes would be difficult if not impossible to engineer. A buck fence is a series of triangles and rails that sit on top of the ground. Since it is usually made of long, thin poles, often with the bark on, a buck fence looks very rustic.

It would be an open invitation for wood chewing if used for a small confinement area. It is most suitable for large mountain pastures.

Polyethylene-Coated Wood

This recent innovation — planks and square wooden posts covered with a polymer resin — is usually available in the traditional horse farm white. The ends of the boards and posts have protective caps to prevent moisture from getting into the wood. This fence requires very little maintenance, is attractive, durable, safe, and highly visible, and deters chewing.

Pipe Fencing

Pipe makes a very strong, safe horse fence, particularly good for pens and runs, and needs little maintenance, especially if rust-proof metal or paint is used. Small-diameter pipe posts must be set in concrete. Usually 2-inch drill stem pipe is used for the top and bottom rails with three or four strands of cable or sucker rod in between. Because of the strength of the pipe the rails can be 10 or 20 feet long, saving on the number of posts that will need to be installed.

Pipe fence offers little flexibility, however. Pipe fence looks very businesslike, but some feel it lacks the aesthetically pleasing look of other types of horse fencing. Materials can be expensive, unless you are located near an oil field where the 2-inch drill stem pipe and sucker rod are available as surplus. A cutting torch, portable welding unit, and someone to operate them are also needed.

Stone

Stone walls make very attractive fences, reminiscent of the British countryside. They are virtually indestructible and also offer a windbreak. They can be used over rocky terrain where posts are difficult to set. The cost of labor to construct them, however, is incredibly high, and unless you have access to stones or rocks on your property, the materials and hauling can be very expensive also.

Concrete

Concrete-formed fences will never rust or rot but they may crack or crumble unless they are engineered and installed properly.

Hedges

Hedge rows are an attractive, natural fence, but in order to be effective they must be about 10 to 15 feet wide and 8 to 10 feet high; otherwise, horses jump or push through them. Hedges take up a lot of pasture space, so are only suitable for very large fields. They require good soil, climate, and moisture and take three years to reach fence size.

ALL WEATHER FENCE, INC

ABOVE:
Which type of fence is best for you? Polyvinyl-chloride or wood?

BELOW:
A buck fence is appropriate for rocky country where digging fence post holes would be difficult.

Pipe and wire cable fence. COURTESY OF TED AND LYNN BROWN

They may also require additional watering, pruning, and fertilizing. Depending on the variety of hedge, horses do eat the bushes, and thus yew and deadly nightshade are to be avoided due to their toxicity.

Woven Wire and V Mesh

This kind of fencing comes in continuous rolls usually 4 to 5 feet tall. The fence is often set 6 to 12 inches off the ground. The openings in the fencing are either squares or diamonds and are about 2 inches by 4 inches and made of 12½ gauge wire. If set right on the ground, it is good for keeping out small animals and children and also for collecting trash on its windward side.

There is an important difference between welded wire and woven wire. Welded wire is spot-welded together at junctions; woven wire is actually tied together with special knots. When a horse crashes into or rubs on a welded wire fence, the welds often break; woven wire joints do not. All wire fences need bracing at the end and corner

posts or the fence will stretch and sag. Especially during shedding season, horses love to rub on this fence, adding to its sagging and bulging problem. Often, to maintain fence shape, electric wire, smooth wire, or boards are used at the top of wire fences to keep horses from leaning over it.

V Mesh is safer than square stock fence. It would be unusual for a mature horse to get its hoof caught in V Mesh, although a foal could. When a shod horse paws at a woven wire or V mesh fence, however, the heels of its shoes can easily be caught and pulled off.

Wire fences should have a galvanized (zinc) coating to delay rusting. The thickness of the coating determines the length of time the fence is able to resist the corrosive effects of the weather.

Chain Link

Chain link is a type of mesh fence with a framework of both horizontal and vertical pipes. Like V mesh, it is also prone to stretching and sagging if abused, but is excellent for keeping out large and small animals and children. It is good for high residential areas and for keeping foals or stallions. As long as the sharp ends at the top and bottom are covered either by cement or pipe it is generally safe. It is not a particularly beautiful fence, but is respectable looking if well maintained. It is very expensive initially but if properly installed, requires little or no maintenance.

Hi-Tensile Wire

Single strands of smooth or twisted barbless wire should be 12½-gauge hi-tensile galvanized wire. Properly installed, it is stretched extremely tight and will withstand 200,000 pounds per square inch. This means not only that it is strong, which is good, but also that if a horse gets tangled in it, the horse will likely come out the loser. Not a highly visible fence, it is most suitable for large pastures or rangelands.

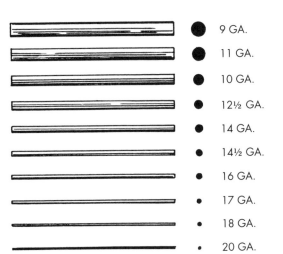

Drawing #25
Wire gauges. Fencing is made from various guages of wire, from very heavy to very light.

9 GA.
11 GA.
10 GA.
12½ GA.
14 GA.
14½ GA.
16 GA.
17 GA.
18 GA.
20 GA.

Vinyl Mesh

Polyethylene mesh fencing is low maintenance if properly installed and is available in a variety of colors. Though it is light to handle — one-fourth the weight of a similar height of wire mesh — it can tend to stretch, flap, and get loose. Depending on the pattern of the material, a vinyl mesh fence can appear less airy and more like a solid fence.

Polyvinylchloride

Often referred to as PVC, this type of material is available in several styles: post and board, post and rail, and post and ribbon. Used in traditional white, it makes a very handsome fence that never needs painting. One drawback, besides its initial price, is the fact that in temperate climates, the freeze and thaw of the ground can cause the posts to shift and the rails to pop out and sometimes shatter. On the plus side, because PVC fencing has elastic action and is highly visible it results in less injury and veterinary bills than many other types of fencing. Another advantage is that horses don't generally chew PVC. Originally, PVC would deteriorate from ultraviolet rays, but today's PVC is highly resistant.

ABOVE:
V-Mesh fence is a safe wire horse fence.
COURTESY OF COUNTRY STORER RANCH.

BELOW:
This entry-way to stalls from an outside paddock is a combination of a durable mesh and tubular metal posts with braces for extra strength and security.

One way of running wire fence with wooden posts.

Ratchet tighteners for a wire fence. COURTESY OF PLEASANT VALLEY FARM

Hi-Tensile Polymer-Coated Wire

This is a flat ribbon type of fence with hi-tensile wire encased in a solid polymer. It comes in white in 1-inch- to 5-inch-wide strips in continuous rolls. Polymer-coated wire is also available.

Rubber Belting

Conveyor belts and specially designed two-ply nylon-reinforced rubber fencing are strong and resilient but unless properly in-stalled can allow persistent horses to escape. Rubber fencing has merit, however, for use as a round pen or as pens for young horses. Because of the give, the posts must be set close together and the belting spaced fairly close also. Rubber fencing invites chewing and colic has resulted when horses ingested the "rubber" and nylon reinforcing fibers. Seal frayed edges by singeing with a hand-held propane torch, then coat with a rubber latex paint or blacktop sealer.

Commercial rubber fencing comes in 1,000-foot rolls weighing approximately 450 pounds. Once properly installed, it is virtual-ly maintenance-free.

Electric

Electric fencing falls into a category all of its own and requires some specific comments. It can be used temporarily by itself, as a means of training horses to stay off other types of fencing, or as a safeguard and chew-ing-deterrent used permanently with other types of fencing.

An electric fence system consists of wire, insulators, posts, and a grounded power source.

Wire. There are basically two types of elec-tric fence wire: solid steel and polypropy-lene — plastic wire with strands of stainless steel woven into it. For horses and managers, plastic has great advantages over steel. If a horse gets tangled in wire, plastic will more easily break while steel will often cut the horse. Plastic fence can be put up or repaired with just a pair of scissors and a simple knot. It is light and flexible and does not need to be stretched tightly as steel wire does.

If you decide to use steel wire for your electric fence, choose galvanized 12-gauge smooth or barbless twisted galvanized cable. Sixteen-gauge or lighter restricts the flow of the current and limits the efficiency of the fence.

Insulators. The traditional insulator used with steel wire is a ceramic insulator but they are getting more difficult to find and are costly. With the advent of plastic fence wire came plastic insulators. You can use either wooden or metal posts with either type of electric wire. Generally posts are set about 8 to 12 feet apart.

Power source. The power source is called a controller, charger, "fencer," transformer, or energizer. It is usually one of three types: battery, plug-in, or solar. Six-volt dry-cell-battery-powered transformers are expensive when their limited life (one to six months), is considered, especially if used steadily or for a large pasture. This battery is not recharge-able and a short will drain it completely. Plug-in electric transformers run off house-hold current at 110 volts. Solar transformers have photovoltaic cells that charge a battery, and these can last three weeks without a sunny day.

Some models emit a pulsating charge, others a steady one, and some can be switched from one to the other. High voltage units are highly resistant to being grounded out by tall grass and weeds.

The power source (controller) should be properly installed. Follow the directions for your particular model, but here are some general guidelines. Some models require a clean dry location where moisture cannot drip or blow onto the unit. Others are de-signed to be weather resistant and can be mounted right on a fencepost; however,

Extended insulators for an electric fence.

An insulated electric fence gate handle.

it is still advisable to protect them by a weatherproof box. The exception to this is the solar fence charger, which must be open to the sun.

If your power unit is inside a building, be sure to install a tube insulator in the wall so you can run the "hot" wire from the unit outside to your fence without it being grounded by the wall. A piece of small-diameter plastic pipe works well.

A solar electric fence charger. COURTESY OF PLEASANT VALLEY FARM

Grounding. The controller unit itself must be grounded. In order for a horse to feel a shock, the current must travel from the controller to the wire to the horse to the ground and back to the controller. (That's why a bird can sit on the fence and not be shocked —it hasn't completed the circuit.)

Most fence problems are caused by improper grounding. The ground rod or tube should be of copper, steel, or galvanized pipe and should be driven 6 to 8 feet into the ground where it reaches permanently wet earth. (If you are in a very dry climate, your ground rod may need to be driven deeper than 8 feet.) This will result in the controller working steadily and efficiently. Connect the ground rod to the controller using number 14 copper wire. A hose clamp works well to attach the wire to the rod. You may

wish to solder the attachment for a positive connection.

Using the fence. You will be able to power from three to twenty miles of electric fence with your controller, depending on the type and model. Just the fact that the transformer is making a clicking sound does not mean the fence is working. Although horses seem to be able to sense when a fence is working and when it is not, few humans have that innate ability. You should purchase a fence tester so that you can check to be sure the current is flowing through the fence without having to touch it! And you may wish to post warning signs for children so they will not be unnecessarily frightened if they happen to touch your fence.

You should formally train your horses to respect an electric fence by putting feed across the fence from them. Once they realize what this new fence means, they will be less likely to run through it.

Gates. Gates can be reinforced with a single hot wire and an insulated gate handle. Design electric gates to be dead when opened, by having the handle disconnect from the side of the gate that is toward the charger. Or you can bypass a gate in an electric fence by running an insulated cable under the ground. For this, use type UF cable (direct bury), 12-gauge 2 wire, or 12-gauge 2 wire with ground, but you only need to use one wire. You could also use extra tall posts on both sides of a gate and run the wire over the top. In such a case, the wire should be at least 8 feet high. Keep all grasses, weeds, bushes, and trees clear of electric fences or they can ground out the entire fence, making it nonfunctional.

Shorts are often caused by faulty wire — that which has become rusty (metal) or frayed (plastic). The electric current will arc and jump across the faulty spot. This causes a burn spot on the wire (weakening plastic wire to the point of breakage) or can even cause fire if combustible material is nearby.

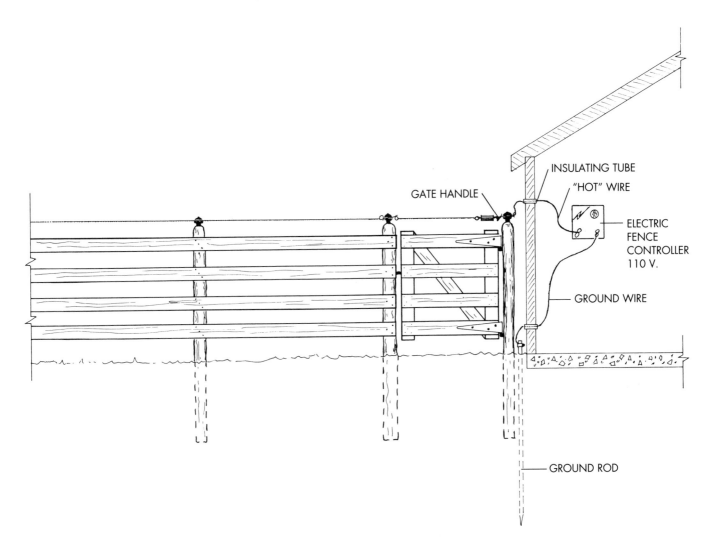

INSULATING TUBE

"HOT" WIRE

GATE HANDLE

ELECTRIC
FENCE
CONTROLLER
110 V.

GROUND WIRE

GROUND ROD

Shorts can also interfere with radio and television reception.

"Riding Fence"

On the large cattle ranches of the west, one ongoing task is checking fence. Although you may not need to saddle up for this job, you should include in your routine a regular inspection of all of your fences. Look for worn or weathered materials, rails that have been broken or dislodged from the poles, electric fence that has shorted, leaning or loose posts, and places where the fence might be sagging and need to be tightened. Pick up the trash that collects along your road fence. Walking the boundaries of your paddocks and pastures every day will also allow

Drawing #26
Electric fence properly installed:

Handle connects on the side of the gate toward the charger so the gate wire (and fence) is dead when unhooked. There is a screw eye on the fence post to hang fence handle when disconnected.

When going through building walls with electric wire, use porcelain tube insulators, pieces of garden hose, small pieces of plastic or rubber tubing to insulate the wire.

Be sure to install a proper ground.

Use electric to fortify or protect your fence, not as a permanent, electric-only fence for horses.

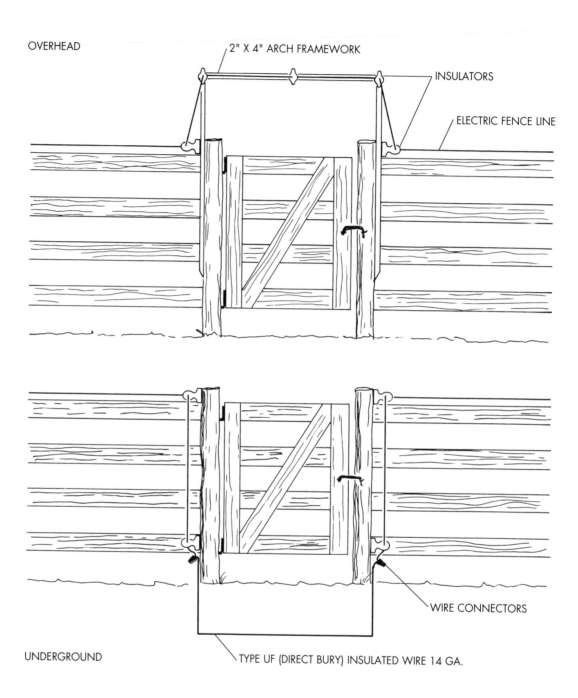

OVERHEAD

2" X 4" ARCH FRAMEWORK

INSULATORS

ELECTRIC FENCE LINE

WIRE CONNECTORS

UNDERGROUND

TYPE UF (DIRECT BURY) INSULATED WIRE 14 GA.

Drawing #27
Continuous electric
gateways.

you to assess whether the land is being overgrazed or if it needs other attention.

Turn-Out Areas

Remember that the smaller the enclosure the greater the chance the horse could get hurt, so, although all fencing must be safe,

be sure to choose the very safest fencing you can for pens, runs, and paddocks. Make sure corners are safe, that waterers and feeders do not protrude with sharp edges or create dangerous spaces where a horse could get caught. Be certain that there are no protruding bolt ends; use round-headed bolts (carriage bolts) whenever possible. Design all gates to be flush with the fence when closed. Roof edges and the corners and

TRIPLE CROWN FENCE

An attractive paddock fence. For the horse's safety, he should not be turned out wearing a halter.

COMPARATIVE COST OF FENCING

Types of fencing	Relative initial cost*
WOOD BOARD FENCE	2.0 X
POST AND RAIL	1.5 X
BUCK FENCE	1.0 X
POLYMER-COATED WOOD	3.0 X
PIPE	2.0 X
STONE	5.0 X
CONCRETE	2.0 X
HEDGES	1.0 X
WOVEN WIRE	2.0 X
WIRE V MESH	2.0 X
CHAIN LINK	4.0 X
VINYL MESH	2.0 X
POLYVINYLCHLORIDE	3.0 X
RUBBER	4.0 X
ELECTRIC	.2 X
HIGH-TENSILE WIRE STRANDS	.2 X
*Per linear foot based on X	

bottom edges of metal sheds are particularly dangerous, and turned-out horses should not have access to them.

There should be absolutely no junk, garbage, or machinery in any area that horses frequent. Check along road fences for litter such as cans, bottles, and items blown out of vehicles. Guy wires for telephone poles, power lines, or antennas should not be located in horse pastures. If you cannot get around this, be sure to tie something on the guy wires so they are more visible or set a pair of wooden posts with a rail between them to shield a horse from the guy wire.

The areas where horses are turned out vary in size, footing, and amount and kind of vegetation present. *Pens* are usually the size of a generous box stall (16 feet by 16 feet) and are meant to be a horse's outdoor living quarters. Horses in pens must be ridden or allowed free exercise in a large area daily. The footing in a pen can be native soil, pea gravel, sand, or bedding. If bedding is used, then the pen must be covered and protected from the weather.

A *run* is usually a long, narrow pen specifically designed for exercise. A 20-foot-by-100-foot run will allow a horse to trot; if you wish to encourage your horse to gallop, you will have to provide about 200 feet and

ABOVE:
A maximum security tie area. Note the horse is tied to a post, not a rail; the level of the tie is at or above the level of the withers; note the length of the tie. A chain is weather resistant and horses do not chew chain as they do ropes.

RIGHT:
A tie area designed for young horses. The wall is solid for safety; the foal is tied to an inner tube rather than to a nonyielding chain.

make sure there is enough room for him to turn around safely at high speed at the end of the run.

A *paddock* can be thought of as a large grassy pen or a small pasture, ranging from one-half to several acres. The grass must be monitored carefully or overgrazing can turn a paddock into a dirt lot in a hurry.

Pastures are improved, well-maintained grazing areas provided mainly for their nutritional value, with the added bonus of exercise. See chapter 12 on pasture management for more information.

Training Facilities

If you can afford the cost and space, consider putting in an outside tying area, a round pen, and an arena.

Tie Area

An outside tying area is a handy training device, allowing you to teach a horse that he must stand relatively still for several hours at a time. This teaches a horse to accept restraint and be patient, and it will be reflected in his attitude about many other activities. The tie area should be very strong and tall, preferably a solid wall. Unless you have unusually large horses, a wooden wall that starts 2 feet off the ground and goes up to 6 feet should suffice. A board fence with spaces between the rails might allow a horse to get its legs caught between them; a solid wall is safer. The place where the lead rope is attached to the tie area should be at the level of the withers or higher so that if the horse does pull back, he will not be able to get very good leverage.

Round Pen

One of the most valuable pens you can have is a safely constructed round pen. Besides providing a good place to turn young horses out for exercise, a round pen is ideal for conducting the following training lessons: restraint, sacking out, longeing, ground driving, saddling, ponying, first rides, and rider longe lessons. In addition, trained horses can be tuned up using work in a continuous circle.

Size. The size and construction of a round pen depend somewhat on its intended uses.

A 66-foot post and
plank round pen
with sand footing.

A "breaking pen" that is built solely for initial gentling and first rides might be 35 feet in diameter with solid 7-foot walls to assure maximum control in unpredictable situations. Solid walls tend to make the horse pay exclusive attention to the trainer. Although this can be an advantage with a snorty bronc, there comes a time when the horse must listen to the trainer despite outside distractions. Some horses have to be retrained when they are taken out of a solid-wall pen and into the real world.

The trainer will be less likely to get a foot hung up when riding in a solid-wall breaking pen than in a post and rail or plank pen. On the other hand, being slammed into a solid wall or getting wedged between a horse and a wall can also be dangerous. Scaling a 7-foot solid wall is difficult for a horse, and next to impossible for a trainer. In addition, the lumber required to construct a solid-wall pen can be cost-prohibitive.

In contrast to a breaking pen, a "training pen," designed for routine longeing, driving, or riding, should be 66 feet in diameter to allow a horse sufficient space for balanced movement. A 66-foot-diameter circle corresponds to the 20-meter circle used in dressage training. Asking a young horse to perform routinely in a circle smaller than this can cause stiffness and stress and can lead to unsoundness. The walls of a training pen are often shorter and more open than those of the breaking pen.

On most horse farms, the training pen gets many more hours of use than does the breaking pen, so if you have to choose, it's probably better to opt for the larger pen. Here are some specifics on a training pen that have worked well for me.

Construction. Once the site for the round pen has been selected, the footing needs to be examined. If necessary, the ground should be graded to ensure a level training surface and proper drainage.

Then proceed to dig your post holes. To measure for the post holes, affix one end

The sand is held in the round pen by fitted railroad ties.

SUPPLY LIST FOR A 66-FOOT TRAINING PEN

28	7" diameter 10-foot pressure-treated posts
2	7"diameter 12-foot pressure-treated posts
113	2" x 8" pine boards, 8 feet long
2	2" x 8" pine boards, 12' long
1	2" x 8" pine board, 10' long
30	6" x 8" x 8' railroad ties (used)
1	230' of ½"-dia. cable with turnbuckle and 4 clamps (2 pieces 115')
1	5½' wide x 5' tall gate
2	10" strap hinges with 16 wood screws (#12 x 1½")
1	Gate hook with screw eye
40#	5" screw shank nails
65	Tons of sand (50 cubic yards)
14	Sacks of dry mix sand and cement
30	Post holes
2	Gallons wood preservative

of a 33-foot rope to the ground at the center of the pen. Walk the rope in a complete circle as a final position check. Decide where the gate will be and mark holes for the gate posts about 6 feet apart. Then mark twenty-eight spots at 7-foot intervals along the cir-cumference of the circle for the rest of the post holes. Posts that are set 7 feet apart, center to center, utilize 8-foot boards with the ends trimmed.

The 10-foot posts are set about 3½ feet into the ground, leaving 6½ feet exposed for the wall. The 12-foot gate posts will extend 8½ feet above ground level. If the posts are tilted outward at an angle 5 to 10 degrees from the vertical, the rider's knees and feet will be less likely to take a beating when an inexperienced horse crowds the rail. Depending on the soil type in your area, you may need to use a cement mix to stabilize the bases of the angled posts.

You should set all of the posts except two, so that a truck and/or tractor can fit inside the pen for dumping the sand and leveling it. In many parts of the country native soil is inappropriate for training surfaces, as it is hard and poorly drained. Four to six inches of sand provide adequate cushion and drainage in most locales and

for most training situations. Sand can be abrasive but, unless it has become compacted, has good shock absorbency. Sixty-five tons of sand result in a 4-to-6-inch cover for a 66-foot-diameter round pen.

Cut both ends of twenty-nine of the 8-foot railroad ties at about a 70-degree angle so that they wedge between the bases of the posts to hold the sand in the pen. Placing the 6-inch side of the ties on the ground allows the 8-inch side to act as the most effective sand barrier. The ties should be wedged in from the inside of the pen and should be flush with the inside surface of the posts.

The boards for the round pen walls can be rough-cut or planed, green or dried, treated or not. The cheapest boards are usually untreated rough-cut wood. Although rough-cut lumber has the advantage of being stronger, it can also be a source of splinters, and it soaks up more paint or preservative than does planed wood. A "2-inch" planed board is actually only about an inch and a half thick. Green lumber can shrink or warp after you have nailed it into place while dried lumber should retain the straightness it had when purchased. Treated lumber resists weather but adds an additional cost.

The 8-foot boards are cut to fit and nailed on the inside of the pen using 5-inch galvanized screw shank nails. The spiral ribbed surface of these nails provides a better grip in the wood than smooth shank nails. Nailing the boards on the inside of the pen not only results in more strength, but in the event a horse leans heavily or falls against the rails, the boards are less likely to pop off. This also keeps the rider's leg from being thumped by every post when trying to control a frightened or bolting animal. Use four or five boards as desired between each set of posts, except for the opening between the two taller gate posts.

To keep the angled walls of the round pen from sagging outward, it is a good idea to reinforce the outside of the pen walls. A ½-inch steel cable with a turnbuckle will provide the required support. One end of each of the 115-foot cables is wrapped around each of the gate posts and affixed with a U-clamp. The cables encircle the outside of the entire round pen about 4 inches from the top, resting in small notches made in each post with a chisel, saw, or ax. They meet across the pen from the gate, where the turnbuckle joins them together. Final tightening is made with the turnbuckle.

The outward pull created on the gateway posts by the cable must be counteract-

ed, or else the gate opening would continue to widen. A railroad tie affixed between the tops of the gate posts can be a stabilizing lintel for the opening. Chiseling a hole in each end of the tie and carving corresponding projections on the tops of the gate posts at least 7 feet high creates a mortise and tenon junction. An alternative lintel could be made with a piece of ½-inch cable and a turnbuckle attached to the tops of the two gateposts. For safety, 2-inch-by-6-inch boards should cover the cable from both sides.

The 5½-foot-wide gate is constructed out of the same dimensional lumber as the round pen walls and is designed to be flush

ABOVE:
Boards are nailed on the inside of the round pen posts, and the walls slope outward at a 5- to 10-degree angle.

BELOW:
To stabilize the sloping walls, a cable encircles the round pen and is fastened with a turnbuckle.

TRIPLE CROWN FENCE

An arena fence designed to prevent injury to rider's legs.

with the inside of the round pen. Ten-inch strap hinges are screwed to the inside of the gate allowing it to swing inward. The hook and screw eye latch is located on the outside of the gate.

If the pen is to be used for turn-out, it may be a good idea to treat the rails with a wood preservative to discourage chewing.

Arena

The size and type of arena you construct will depend on the type of riding you plan to do. See the chart on the following page for some suggested dimensions for various activities.

Fencing. If your arena fence is at least 6 feet tall, it will discourage horses from putting

their heads over the rail as they are turning near the fence. The fencing should be very strong if you plan to ride young horses. Often dressage rings have no official exterior fencing but just an 18-inch-high visual barrier. The perimeters of dressage rings are commonly marked by cones, plastic chains or ropes, or with PVC pipe and cinder blocks.

The shape of your arena will depend on your training goals. Rectangles allow you to ride your horse deep into the corners and teach him to bend. Oval arenas or rectangles with rounded edges are more appropriate for driving and jumping and are easier to disc and harrow. Gates should be flush on the inside of the arena and the latch should be operable from horseback.

ARENA DIMENSIONS

Use	Size Required
DRESSAGE (small size)	20 meters x 40 meters 66 feet x 132 feet.
DRESSAGE (large size)	20 meters x 60 meters 66 feet x 198 feet.
CALF ROPING	100 feet x 300 feet.
TEAM ROPING	150 feet x 300 feet.
PLEASURE RIDING	100 feet x 200 feet.
BARREL RACING	150 feet x 260 feet.
JUMPING	150 feet x 300 feet. (Depending on number and type of jumps and type of course)

Construction. All arenas should be either crowned at the center or sloped gradually from one side to the other. Choose a site that requires minimal excavating. Bulldozing and grading are very expensive and the less earth that has to be moved, the cheaper the final project will be. While the heavy equipment is there you may need to install some ditches to divert surrounding drainage away from your arena. After excavation, the arena site will have to settle for six to twelve months, then be leveled before you add any footing material on top of the base.

Footing must be well drained and of appropriate cushion. The type of footing you choose will depend on your climate, whether the arena is indoor or outdoor, and what type of activity you participate in. Jumpers require cushion without excessive depth. Speed events require a firm footing such as a mixture including stone dust. Reining horses do best on a firm base with a slightly slick top of sandy loam. Dressage horses work well on a resilient footing without excessive depth, such as some of the processed wood products.

One of the most common ways of improving native soil is to disc sand and/or

sawdust in with the dirt. This will lighten and loosen the soil and increase its drainage while adding to its cushion. It takes about 250 tons of sand to provide a 4-inch cover in a 100-foot by 200-foot arena. If you are trying to firm up the footing add stone dust, a little at a time, until you reach the desired consistency. The total footing will probably consist of no more than 10 percent stone dust.

There are processed footings available that can be spread over a firm arena base, but if you are investing in one of these, it might be better to use it in an indoor arena. Tan bark, hardwood fiber, and wood chip products tend to freeze later and thaw sooner than the surrounding ground. They don't need to be disced, just lightly harrowed. Besides the high expense of the footing itself, however, processed wood fiber footing requires a well-engineered drainage system in order for it to work at its optimum.

It may frequently be necessary to disc, harrow, and in some cases water the arena footing to keep it in an ideal state for training. After a period of time the "track" on which you ride will become either rock-hard or grooved as in a slot-car game, hampering your horse's movement.

Various means of controlling dust have been tried, but be sure to think of the consequences before you implement them. Adding oil, especially motor vehicle waste oil, to the footing may keep dust down but petroleum products are thought to be harmful to a horse's health, and if used in an indoor arena, the miserable residue settles on your clothes and tack and in equine and human lungs. Water keeps dust down but evaporates quickly and is costly to apply.

To keep the footing from freezing, some indoor show arenas use rock salt or calcium chloride, but this can be drying to hooves. And if you use a salt on an outdoor arena, what would the run-off do to the vegetation near your arena? See chapter 12 on pasture management and chapter 13 on water to find out the answer.

SECTION THREE
MANAGEMENT

··

I WAS VISITING A NEW HORSE FARM THAT WAS still under construction. Although the barn was not completed, the stalls were being used, out of necessity. As I talked with the farm manager, another visitor drove in the lane. The driver got out of his truck, followed by a small yapping dog that immediately chased one of the farm's cats right into a stall occupied by a horse. The startled horse reared and broke the yet-uncovered light bulb over his stall.

The horse received a cut on his poll, so the manager immediately took him out of the stall to remove the shards of glass and treat the wound. What none of us realized was that the hot filament from the light bulb had dropped into the stall and ignited the fresh straw bedding. The already agitated horse alerted us to the smoke. The manager grabbed a fire extinguisher, which was located conveniently by the barn door, and quickly doused the straw. He then switched off the power to the barn until he could take the time to repair the broken fixture.

The canine cause of this chain reaction of mishaps was still running loose, paying little attention to his owner's calls to come. The visitor said with a forced laugh, "If I start the truck as if I'm leaving, he'll come." The manager's manner made it clear he was not joking when he said, "There is no 'as if' about it — you're leaving."

Sure enough, the dog hopped into the truck when he heard the engine. As the driver pulled out of the farm I noticed his peeved expression, and I wondered what version he would tell. Meanwhile, the manager was making plans to post "ABSOLUTELY NO DOGS ALLOWED" signs conspicuously at the farm's entrance.

CHAPTER TWELVE

Pasture and Hay Field Management

WHEN PLANNING PASTURE IMPROVEMENTS, RENOvations, and routine management plans, confer with your local agricultural extension agent for advice on soil testing, plant varieties, fertilizer and irrigation needs, weed control, and a timetable for planting and harvesting.

To make some of these decisions, you will need to consider various aspects of the climate such as the temperature, growing season, and grazing season as well as the expected time and amount of average precipitation.

Soil Testing

One of the first places to start is to have the soil tested to determine if it needs mechanical or nutritional improvement. Ideally your soil should be fine-textured and fairly moist, well aerated, not heavily compacted, and composed of a mixture of sand, loam, and organic matter. Such a soil is usually friable: that is, if you grab a handful of the soil, squeeze it tightly into a ball, and then try to break the ball, it should crumble.

If you are lucky your land will have 3 to 5 feet of topsoil. You will want to protect your topsoil by avoiding intensive agriculture and minimizing erosion by wind and water. Avoid over-grazing, seed and mulch bare spots of soil, and plant trees and shrubs to minimize wind erosion.

To see how well drained your soil is you can perform a simple percolation test. Dig a hole 3 feet deep (about 12 inches wide). Add 2 inches of gravel to the bottom of the hole. Fill the hole one-third full of water and keep adding water to maintain it at that level for four hours. Then let the water level go down to 6 inches above the gravel. Measure the drop in the water every thirty minutes. Multiply the average times two for the percolation rate per hour. A rate lower than 2½ inches per hour is undesirable.

In order to establish if your soil needs nutritional supplementation or chemical balancing, you will need to submit a soil sample to your local agricultural extension agent. Soil sampling procedures are fairly standard. First you will need to determine the various distinct areas of soil on your property by noting the following factors: difference in texture (sand, silt, clay, etc.), color, slope, degree of erosion, drainage, and past management (fertilization, cul-

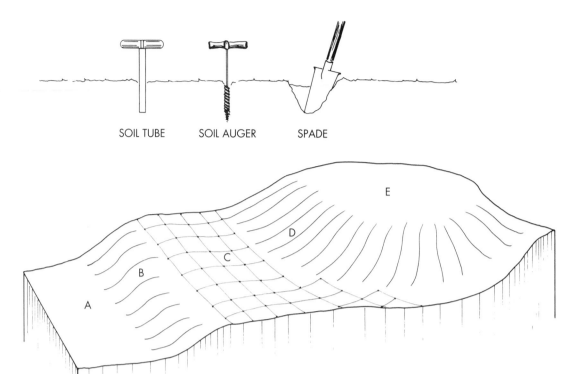

SOIL TUBE   SOIL AUGER   SPADE

Drawing #28

Soil sampling.

Take soil samples using either a soil tube, soil auger, or spade. Divide your land into soil areas (ABCDE) and sample each area separately. Take 20-30 samples from each area in a systematic manner. (See area C)

tivating, grazing, etc.). Make a map of your land and draw in these areas and number them.

Next take samples to cultivation depth (usually 8 to 12 inches) from each area. Use a stainless steel or plastic soil tube or auger or a clean spade as rusty tools can contaminate samples with iron. Systematically take twenty to thirty samples from every uniform area. Avoid very unusual spots (or sample them separately) such as waterways, hedge rows, piles of decomposing organic matter, etc.

Mix all samples thoroughly in a plastic bucket: galvanized steel or brass containers will contaminate the sample with zinc, whereas rusty steel may contaminate it with iron. After mixing, take out enough soil to fill the sample container. Air-dry the sample by spreading the soil in a thin layer in a clean, room-temperature location. Label the sample container with your name, address, and the sample number (corresponding to the area on your map). Fill out the soil information form completely. Pack the sample container.

The results of the soil test will yield some valuable information. Routine tests are usually made for pH, organic matter, ni-

trate-nitrogen, phosphorus, lime estimates, micro-nutrients, soluble salts, and texture. This generally gives an adequate soil profile unless a special problem is suspected.

pH indicates soil acidity. Neutral soils have a pH value of 7.0. Sour (acid) soils have lower values and sweet (alkaline) soils have higher values. If your soil is too acid, it will be recommended you add lime to sweeten it.

The proper addition of nitrogen to deficient soils can double hay and pasture yield. Such an increase could result if an established, nitrogen-deficient pasture received 50 to 60 pounds of nitrogen per acre at the beginning of the growing season and then twice more during the season at six-week intervals.

Phosphorus, especially important for clover and other legumes, requires a season to take effect. If your soil is deficient in phosphorus, therefore, you will want to apply it in the fall for spring growth.

Micro-nutrients such as zinc, iron, potassium, calcium, magnesium, copper, and manganese are important for plant emergence and vigorous seedling growth.

A routine soil test will also indicate the salt and sodium content of your soil. A high level of soluble salts will greatly limit your crop yield. Saline soils cannot be corrected by the addition of chemicals, conditioners, or fertilizers. The salts must be removed by leaching or flushing with water and possibly replacing the sodium with calcium. The irrigation water used for flushing must be evaluated. If it is of poor quality (also high in salts) it will contribute to the problem.

Making Plans

Whether you plan to use your pasture for year-round grazing or whether you have decided that it would be more profitable to take one cutting of hay before allowing your horses to graze, the principles of raising hay will give you a good idea of how to improve a pasture.

You may be interested in developing temporary pastures, permanent pastures, or both. A temporary pasture is intended for one to two years of vigorous year-round grazing. Because the plants are never allowed to develop deep roots they can be more susceptible to damage from hooves. Cool season plants such as rye, wheat, alfalfa, and clover work well for temporary pastures. Of course, the investment of labor to initiate such pastures is high in relation to the number of years of use.

Permanent pastures are those planned to be used for ten or more years before major renovation. Along with a well established base of native plants they may have an introduced population of warm-season perennial grasses. Such a pasture has a well developed root structure that tolerates temperature and moisture stress and resists mechanical damage and erosion.

Improving Your Pasture or Managing Your Hay Field

The story of the hay in your horse's feeder may have started two to three years ago when a field was prepared and seeded. Often hay is used in rotation with other crops such as corn or barley. While such grains deplete

Because this pasture was not overgrazed in the summer and fall, the grasses catch the winter snow and provide winter grazing and an exercise area for this broodmare.

the soil of its nitrogen, legume hays such as alfalfa restore this necessary element back to the soil. Because of this rejuvenating effect, along with the fact that premium hay is a good cash crop, alfalfa is a very popular hay to raise. Furthermore, a large, well established and properly managed grass or mix pasture can produce a good crop for first-cut hay, and then be used again for grazing.

To establish either a hay field or a pasture, you would follow a similar procedure.

Preparing the soil. Fields can be worked using conventional methods or alternative tilling. With conventional tilling, the field is cultivated deep with a mold-board plow, disc, chisel, or rotary tiller, and then can be further worked with a mulcher to mix and aerate compacted soil. Following this, the field is floated (leveled) with a disc and/or harrow to smooth out the rough spots and to prepare it for seeding. Hay and pasture plants do best in a fine, firm, clod-free seed bed, not an overly soft or clay-pan field.

Due to the high cost of fuel, labor, and equipment, alternative methods of tillage have been used with success. Minimal tillage might consist of a light discing and harrowing, resulting in less compaction due to less traffic. If you do not till, erosion due to wind or water is greatly reduced but seeds must then be drilled. You may find it best to hire a custom farmer with a renovator or a cultimulcher to work your field. He can cultivate, chop, aerate, seed, and fertilize all in one pass.

Adjusting pH. In some areas, you may need to add limestone to reduce soil acidity and thereby increase yield. A low pH means an acid soil, which may interfere with a plant's ability to absorb minerals. Lime, which is a base, will raise the pH (decrease the acidity). It may need to be added every two to twenty years, depending on leaching. Adding lime is especially important where there is 25 inches or more of rain per year, which can leach lime and plant foods out of soil and leave it acid. Lime can be added on top (even over snow) but tilling it in is better because it becomes more accessible to the plant roots. In contrast to lime, nitrogen fertilizers can increase soil acidity.

Pastures benefit from fertilizer applied at a time when the traffic will not hurt the land.

Fertilizing. If a soil test indicates the need for additional nutrients, fertilizer containing specific ratios of nitrogen, potassium, and/or phosphorus can be added. Fertilizing gets seedlings off to a vigorous start, assures consistently higher yields, and helps plants withstand stresses from insects and winter kill. Nitrogen increases plant growth. A productive, established pasture may require 150 to 180 pounds per acre of nitrogen per year. Remember that legumes create ("fix") their own nitrogen in their root nodules so they require less nitrogen application than do grasses.

Phosphorus is essential for good root development, and phosphates sweeten grass so that horses will tend to eat less palatable grasses. Potassium, meanwhile increases a plant's resistance to drought and disease.

Applying the right fertilizer at the right rate and at the right time increases plant yield, increases a plant's water use efficiency, and decreases weed problems by making the desired plants so vigorous that the weeds cannot get established. Fertilizer can be broadcast and left to dissolve or it can be tilled into the soil. Usually fertilizer is applied to give seedlings a good start and then once more during the growing year.

Seeding. When selecting hay seed, choose the type of hay you wish to grow as well as the variety of hay that does best in your local area. Extension agents have results of field trials that test varieties and mixes for such things as yield, resistance to drought, and the ability to withstand root rot, wilt, and winter-kill. The quality of the seed is dictated by purity (very low weed and other seed content) and germination vitality.

Hay seeds are drilled into the field to a depth of about ¼ to ½ inch for ideal soils and no more than ¾ to 1 inch for sandy soils. Broadcasting seed on top of the soil requires 50 percent more seed than drilling.

Although an alfalfa hay field usually remains productive for about five to six years, and grass fields often longer, not all of those years are equally productive. The key to getting a new field established is taking advantage of winter snows and spring rains by seeding in the fall.

Most often, in order to get a stand of hay or pasture growing, a nurse crop (also called a cover crop or companion crop) is planted along with the hay seed. A nurse crop, such as oats, will emerge ahead of the more vulnerable hay seedlings. While the

oat plants tower over the developing hay seedlings and protect them from the rays of the sun, the root structure of the oats adds to soil stability. The nurse crop is harvested the summer following seeding, at which time the hay is ready for the sun. Often, however, the hay or pasture doesn't grow vigorously enough for grazing or a hay crop until the second year.

Varieties. Grasses such as bromegrass, bluegrass, timothy, rye grass, tall fescue, wheatgrass, and orchardgrass are usually hardy perennials that decrease in digestible nutrients as they mature. Tall fescue is widely adapted, has good tolerance to wet, dry, or alkaline soil conditions, and can withstand a lot of traffic. Orchardgrass produces excellent-quality forage but will not tolerate drought, wet, or alkaline soil. If irrigation water is not available for the entire season, smooth bromegrass and wheatgrass can be used. Bermudagrass (in the south) and tall fescue (in the north and south) resist trampling so are good varieties for "exercise" pastures.

Legumes such as alfalfa, clover, and birdsfoot trefoil are more nutritious and don't require nitrogen fertilizer, but they are less hardy and impractical when used alone for pasture. With pasture mixes, plants mature at different times and exhibit a range of abilities to withstand stresses such as drought, flood, heat, and cold. If a blend of grasses and alfalfa (or other legume) is preferred, the legumes should not exceed 25 percent of the mix.

Irrigation. Heavy irrigation at longer intervals develops better root systems and hardier pasture than do more frequent light waterings. Sandy soils have lower water-holding capacities and therefore require more frequent irrigation, while clay soils are not porous, hold water longer, and require less frequent irrigation. Loam soils generally have water-holding capacities intermediate to sandy and clay soils.

Most pasture and hay fields need somewhere around 24 to 36 inches of water per year, so you may need to supplement your natural water supply to get a maximum yield from your pastures. Irrigation equipment can be expensive and is labor intensive but the yield per acre of alfalfa grown with irrigation is often twice that of alfalfa grown without it.

In order to consider using irrigation, you must have access to a large water source and in some areas "water rights" from a water use agency. You can irrigate by surface flooding, furrows, or sprinklers. Surface flooding is not good for rolling pastures; the land must be leveled and graded with a slope of .1 to .4 feet of grade per 100 feet of length. Since standing water (for more than

Your pastures or hay fields may require irrigation for optimum growth.

twenty-four hours) encourages mosquito and horsefly breeding, irrigation runs (furrows) should be designed and regulated so they can be irrigated and drained efficiently. Sprinklers are difficult to manage in areas of high wind and lose a lot of water to evaporation, so they are best when used in areas of high humidity. Sprinkler systems are the most expensive irrigation method initially but require less labor to operate.

The moisture that is required to grow good hay can also contribute to its demise. Although hay grows quickly in rainy country, it may be difficult to find a gap in the weather pattern big enough to allow harvesting and baling. Dry, sunny regions with adequate irrigation water are great hay-growing areas.

Weed Control

A weed is sometimes defined as a plant out of place. Not all weeds are bad — some are fine additions to a horse's diet, being palatable and containing higher crude protein than grasses and higher trace elements than legumes. Such "good weeds" include dandelion, buckhorn plantain, chicory, comfrey, kochia, and lamb's-quarters.

Some plants, while not "bad," compete too vigorously with desirable grasses and are either unpalatable or of poor nutritional quality. Examples are such things as downy bromegrass (cheat grass), ironweed, horseweed, and willows.

Other plants, while not poisonous, are considered "bad" because of a harmful "mechanical" effect they have on horses. This category would include such plants as foxtail, thistle, and others that cause lacerations and sores in the mouth.

Still other plants contain toxic components that make them poisonous to horses, and these plants should be eradicated in any pasture or hay field. The list of poisonous plants will vary according to your locale, so check with your agricultural extension agent. Some examples are: locoweed, yew, buttercup, poison and spotted water hemlock, deadly nightshade, curly dock, horse nettle, stagger grass, atamasco lily, mustard, hairy vetch, Mexican poppy, hemp dogbane, milkweed, perilla mint, chokecherry, and acorns.

The best defense against weeds is establishing a good, vigorous stand of grasses and legumes that can compete with them. Mowing can impair weed growth by removing the larger-leafed, more vigorous weeds that shade the developing desirable plants. If the plants are cut before the seed pods are mature it can also prevent the start of a new cycle of weeds. Most fields tend to clean themselves of weeds with strategically timed pasture mowing or after the first cutting of hay during the second year.

Burning eliminates mature vegetation and seeds and kills some parasites, but it can be a safety hazard. It is usually done in the spring. Check to see if your local ordinances require a burning permit.

Using Herbicides. It is difficult to control weeds chemically in a hay field or pasture, as many of the herbicides would kill the desirable plants as well as the weeds. You will need to choose a selective herbicide specifically for your situation. Try to control your weeds by other means, but if you must use herbicides carefully monitor every step or you can end up with a much greater problem than a few weeds.

Points to remember when using herbicides:

● Check to see if you need an applicator's license.

● Read directions carefully and mix according to directions.

● Don't think that if a little is good, a lot is better: you could do permanent damage to your land.

● Remove all animals from the area and leave them off as long as directions indicate.

● Be aware of wind drift as you apply —you could inadvertently kill your newly planted fruit trees or destroy your garden.

● Beware of poisoning streams and other water sources.

● Wear a respirator when applying herbicides.

● Keep your dogs and cats confined for a few days after application.

Insect Control

You may also need to deal with diseases and insects. Bacterial, viral, and fungal diseases and some insects will affect hay yield and can kill the entire field. The selection of the appropriate variety of hay or grass for the local growing conditions is often the best preventative. Genetic research has provided us with varieties that are resistant to specific diseases and pests.

Particularly in the Southwest, second-cut or later alfalfa hay may contain blister beetles, which can be lethal to horses. The toxin in these insects, cantharidin, is so deadly that just a few beetles, dead or alive, can kill a horse. If you live in an area where blister beetle poisoning has been reported, confer with your county agent for assistance in identification of the beetles and purchasing certified blister-beetle-free hay.

Predator insects such as wasps, praying mantises, and ladybugs have been used successfully to kill aphids and grubs.

Rodent and Snake Control

Rodents — including mice, gophers, rabbits, ground squirrels, groundhogs, and others — can present problems. First, their holes can pose a threat to your horses as they gallop across the pasture. In addition, rodents can carry dangerous diseases, and a large population can put a real dent in the forage production of your field. I feel it is impossible and undesirable to try to eradicate rodents, but you can keep them under control by keeping a couple of cats and dogs on your acreage and allowing them access to the problem fields. You should also encourage natural predators of rodents such as hawks, owls, coyotes, and snakes.

Be sure you know your snakes. On more than one occasion people have killed harmless and helpful snakes, thinking they were rattlers. Some snakes are natural predators of both rodents and rattlers. Seek out a herpetologist in your area for help in identification.

Pasture Management

A good pasture manager respects, appreciates, and cares for the land. To apply John F. Kennedy's stirring words to pasture management, "Ask not what your land can do for you but what you can do for your land." Do not eke out the last iota of nutrition that the land can possibly offer. Leave some reserve so the pasture can rejuvenate. Protect the land from the damaging effect of overgrazing by horses.

Horses are wasteful and gluttonous and their hooves can be very damaging to the land. A horse will eat, trample, or damage at least 1,000 pounds of air-dry forage per month. This translates to the following carrying capacity range: Productive irrigated pasture may hold two horses per acre per month during the growing season, while 30 to 60 acres of dry rangeland may be required to support a single horse.

Horses go for the young plant growth and succulent roots, letting weeds go to seed and mature plants go to waste. They defecate in certain areas and then will never consider eating the plants growing there unless forced to by starvation.

In addition, during wet periods (natural and irrigation) especially, the hooves of horses ruin root structure and can turn a field into a sea of mud and then a plain of dirt. Horses left on a pasture too long paw to reach tender roots, thereby destroying a plant's ability to rejuvenate after being grazed or cut. Horses should therefore be put on pasture or hay fields when the growth is optimum — about 4 to 6 inches high — and then the plant growth in the field should be closely monitored. Remove the horses when 50 percent of the forage has been ingested or damaged.

If the horses' grazing rate is greater than the field's ability to regrow, they should be put on another pasture so the grazed field can rest for a minimum of four weeks. This concept, rotational grazing, means that land must be divided up into more pieces requiring more fences, waterers, and labor. It does, however, result in a higher stocking rate per acre of land. For example, one acre per horse could be sufficient if horses were rotated between three pastures or between two pastures and a set of holding pens.

Continuous grazing is typically practiced on farms or ranches with very large pastures. This minimizes amount of fencing and waterers, but it can result in seasonal forage shortages and areas that are permanently contaminated with feces and permanently overgrazed. Moving the location of salt and water in such pastures will minimize spots of overgrazing. Continuous grazing yields a low stocking rate — one horse requires two to three times as many acres, or more, for continuous grazing as it would for rotational grazing.

Horses can be mixed or rotated with other livestock to maximize the use of the pasture. If you run cattle and horses together, you run the risk of aggressive horses chasing calves or of horned cattle going after meek horses. If you let cattle rotate with horses in a pasture, however, they may clean up some of the mature grasses left behind by the horses. Since horses and cattle have different parasites, the life cycles of horse parasites will be broken during the time the cattle are in the pasture. Sheep, on the other hand, which tend to eat the center of a plant and leave the tall, tough outer leaves, don't really contribute to the health of a horse pasture.

Pasture horses will have to be fed hay and possibly grain during the late fall, winter, and early spring months. How you choose to do this will depend on how many horses you are feeding. Remember, horses will fight at feeding time, so if you have personality conflicts within your herd, or great numbers of horses, you will need to devise a way of separating horses until each gets its fair share of feed. This usually requires you to put each horse in a separate stall at feeding time, which defeats some of the labor-savings bonus of pasture management.

If you are feeding hay alone, spread it out in the pasture in an open space (provided it is not a windy spot). Place the piles very far apart with several more piles than the number of horses you are feeding. If you need to feed grain or more closely regulate the feeding of hay, I suggest you construct several small pens in which you can separate horses.

When to Cut Hay

The first growth of the second year may be quite weedy and not the best feed for horses, either as hay or pasture. Removing the first cut, baling it, and using it for cattle hay is a good choice. The second cut from the second year will probably be the first crop suitable for horses. The third year marks the beginning of the prime years for an alfalfa or

NUTRITIONAL CONTENT OF HAY

HAY VARIETIES	DE KCAL/LB	TDN %	CRUDE FIBER %	CRUDE PROTEIN %	CA %	P %
ALFALFA	980	50	28	17	.6-2.0	.2
BIRDSFOOT TREFOIL	900	45	27	14	1.5	.2
RED CLOVER	855	44	27	13	.6-2.0	.2
ORCHARDGRASS	820	42	32	9	.25	.25
TIMOTHY	900	45	28	8	.4	.2
BROME	980	49	36	6	.3	.2

alfalfa-mix field. After five years, due to the death of some of the alfalfa plants, the field will have an uneven growth pattern and decreased yield. With a mixed field, the grasses gradually take over and by four to five years, grass will dominate the field.

When to cut hay is critical. Usually it is determined by plant maturity, but other methods involve evaluating crown regrowth after first cut and simply using predetermined calendar dates. No matter which method is used, one eye is always kept on the weather. Hay makers hope for dry but not overly hot days when the hay is in the windrows. Extreme heat or wind can result in dry, brittle hay. Rain or damp weather prevents the hay from drying thoroughly and usually results in bleached or moldy hay.

Using plant maturity as the guide, there exists a trade-off between maximum yield and maximum quality. The premium hay grower chooses the optimum time when the plants are at their nutritive peak. Leaves contain the most protein. Young, immature plants have a high leaf-to-stem ratio so are generally high in protein and low in fiber, resulting in excellent hay but fewer bales per acre. Mature plants, with a low leaf-to-stem ratio, have a lower protein content and higher fiber content. Although this results in a greater number of bales per acre, the bales are of lesser quality.

Legumes, such as alfalfa, should be cut

when the first flower appears in the field: that is, the first flower on a representative plant in the field, not an odd plant along a ditch or field edge. Another way to gauge cutting time is before 1/10 bloom, that is when one out of ten buds have bloomed on the plants. On very large operations, cutting is started at mid-to late bud stage so that cutting will be complete by mid-bloom at the very latest.

Since most grass fields are cut only once, the farmer often waits until the plants are very tall and seedheads are mature. This results in a high yield and a safe roughage, but one with very low nutritive value. Ideally, grasses should be cut at the boot stage, when the seed heads are just emerging from the stem. The emerging head will be short, compact, and resilient, not three inches long, dry, fuzzy, and shedding seeds.

Mixed hays, such as grass-alfalfa, are cut using the maturity of the alfalfa plants as a guide. Each day a plant stands after first flowering or past the boot stage, crude fiber increases and crude protein decreases by 1/2 percent per day.

Using bloom as the sole indicator of plant maturity can be misleading in some situations, as bloom is affected by moisture, clouds, temperature, and the stage of the plants at the previous cutting. If the first cutting was mowed at the bud stage, for example, and adequate moisture was available for regrowth, then the field could be cut every 35 to 40 days after the first cutting. Using such guidelines helps to ensure that there will be three cuttings. Hay cut early is usually of high quality and is followed by a fast regrowth and decent second- and third-cut yields. If a field is cut three times, approximately 45 percent of the year's yield will be in the first cut, 30 percent in the second cut, and 25 percent in the third cut.

Horsemen are very opinionated on which cutting is the best to buy. Although there are some differences in the cuttings, the quality of the hay is much more important than the cutting. From a nutritional standpoint, all cuttings can result in prime horse hay. With alfalfa, there will be some variation in protein content between cuttings. Although first-cut alfalfa hay is reputed to have large tough stems, this is only true if the hay was too mature when cut. If first-cut hay is mowed at the pre-bloom stage, the stems will not be coarse and the nutritive value will be high. Weeds do tend to appear, however, in first-cut hay.

Second-cut alfalfa hay is usually the fastest growing because it develops during the hottest part of the season, and it usually has more stem in relation to leaf. Of all cuttings, second cut tends to be the lowest in crude protein, but its 16 percent average is adequate for all classes of horses.

Third (and later) -cut alfalfa develops a higher leaf-to-stem ratio because of slower growth during the cool part of the season. Therefore, third-cut hay will usually have the highest nutritive value. Horses that are not accustomed to a good, leafy hay may experience flatulent (gaseous) colic or a loose stool.

Mixed hays from all cuttings will have similar nutritional values except that with a grass/alfalfa mix, the first cutting will contain a larger proportion of grasses than will the other cuttings.

Most hay today is mowed, conditioned (stems crimped so they will dry faster), and put in a windrow all in one operation. This results in less manipulation of the hay and less leaf breakage and loss. The hay dries in the windrow until the moisture is out of the stem. The level of dryness can be determined by twisting a handful of the hay. If the stems pop as they break the moisture content is about right for baling. Scraping the green covering off a stem will also reveal if the stem is still wet.

Raking or turning the windrow rolls the hay on the bottom of the pile to the top. This may be necessary in humid climates, if hay has been rained on, or if the stand was unusually dense and the windrows are heavy. Raking will facilitate further drying but may contribute to leaf loss. It is essential that raking be done when the hay has adequate moisture, such as with an early dew, which will prevent leaf shatter and loss.

Baling. Once the hay in the windrow is determined to be at the appropriate moisture level, the hay should be baled with the aid of the morning dew to help hold the leaves on the stems. This may require the hay grower to get up at 3 a.m. and bale for the few hours when baling is optimum. Baling throughout the heat of the day simply does not result in good-quality horse hay, in most situations.

Bale size is dictated, for the most part, by the bale wagon being used, with the currently popular wagon requiring a 40-inch-long bale weighing approximately 65 to 70 pounds. The tightness of the bale can be adjusted. Tight bales handle well, stack well, and shed weather better. A too-dry bale must be baled tight in order to retain its leaves but too-wet hay that is baled tight will result in heating and molding.

Bales are generally left in the field for a few days to cure or sweat, particularly if there was adequate dew on the hay during baling. Often you have to gather the bales because rain is in the forecast or because you need to irrigate the next cutting. Stacking today is generally done with automated bale wagons, resulting in tight, stable stacks with staggered joints. A tall stack results in fewer

HAY VARIETIES AND CHARACTERISTICS

HAY VARIETIES	WHEN COMMONLY CUT	POSITIVE ATTRIBUTES	POTENTIAL PROBLEMS
ALFALFA	first flower	High-quality protein especially for growth and generally a desirable calcium to phosphorus ratio, highly palatable	Needs well-drained soil; will shatter if too dry; can contain too much crude protein for some classes of horses, possibility of blister beetles; excess calcium to phosphorus
BIRDSFOOT TREFOIL	early bloom	Does well in poorly drained soils	Low yield; may have lower palatability
RED CLOVER	early to mid-bloom	Does well in poorly drained soils, high-quality protein	Difficult to put up well, notoriously dusty and possible toxicity from mold
ORCHARDGRASS	boot	Early start, high yield, safe feed	Can get tough and unpalatable after early bloom
TIMOTHY	boot	Does well in poorly drained soils, safe for idle adult	Not drought resistant; when only hay fed, not enough energy for working horse and marginal in crude protein, calcium, and phosphorus for working horse
BROME	early to mid-bloom	Drought resistant, high yield	May be unpalatable if too mature and fed alone; low in protein, calcium and phosphorus

top and bottom bales, the ones commonly lost to weathering and ground moisture. Side bales generally do not get drenched during a rain, so they dry out adequately. The middle bales are protected.

If the bales contain too much moisture, they can ferment and create heat. The heat can be great enough to result in spontaneous combustion, causing an entire stack to catch fire. The internal temperature of a bale can be checked by simply cutting the strings and passing the hand between some flakes. Any warmth should be noted since heat makes undesirable changes in the carbohydrates in the hay.

Good Hay

Good-quality hay should be leafy, fine-stemmed, and adequately but not overly dry. Since two-thirds of the plant nutrients are in the leaves, the leaf-to-stem ratio should be high. The hay should not be brittle but instead soft to the touch, with little shattering of the leaves, since lost leaves mean lost nutrition. There should be no excessive moisture that could cause overheating and spoilage.

Good-quality hay should be free of mold, dust, and weeds and have a bright green color and a fresh smell. In some instances, however, placing too much emphasis on color may be misleading in hay selection. Although the bright green color indicates a high vitamin A (betacarotene) content, some hays might be somewhat pale due to bleaching yet still of good quality. Bleaching is caused by the interaction of dew or other moisture, the rays of the sun, and high ambient temperatures. Brown hay, however, indicates a loss of nutrients due to excess water or heat damage and should be avoided.

Dusty, moldy, or musty-smelling hay is not suitable for horses. Not only is it unpalatable, but it can contribute to respiratory diseases. Moldy hay can also be toxic to horses and may cause colic or abortion. Bales should not contain undesirable objects or noxious weeds. Check for sticks, wire,

One way to control parasites on pasture is to collect manure on a regular basis with a pasture sweeper.

Other Tasks

Mowing. In between grazing periods, it may be necessary to mow the ungrazed portions of the pasture to encourage regrowth and to discourage the plants from getting too mature and going to seed. This may only need to be done one or two times per year but it results in a higher-quality pasture. Remember, just as in hay production, there is an inverse relationship between yield and quality. Usually clipping the grasses to three inches is sufficient.

Harrowing. The principle behind harrowing a pasture between grazing periods is that it evens out rough spots and spreads manure clumps. If you are in a dry sunny climate, this is a good practice as it kills the parasite eggs in the manure. In humid climates, however, harrowing the manure in pastures just spreads the parasite eggs over a larger area while still allowing them to be viable, and so in effect increases a horse's chances of reinfestation. In such a situation, collecting manure and composting it would be best. See chapter 14 on sanitation.

Pasture sweeping. As a means of removing manure and greatly decreasing the potential for parasite infestation, pasture sweepers are helpful. See chapter 10 on farm equipment.

blister beetles, poisonous plants, thistle, or plants with barbed awns such as foxtail or cheat grass.

Purchasing hay. Since the nutritive quality of hay can vary so greatly, it is best to test hay before a large purchase, especially if it is to be used for young or lactating horses. Your extension agent will instruct you on sampling techniques and the test results will reveal crude protein, fiber, energy, and mineral content.

🐎 CHAPTER THIRTEEN

Water

....

GOOD QUALITY WATER IS OF VITAL IMPORTANCE TO the health and well-being of you and your horses, as well as for the proper functioning of your acreage. Become familiar with the source of your water. Does it come from a city or municipal source or is it from your own well? If the latter, where is the watershed that feeds your well and other surface water on your land? Are there waste dumps or land-fills nearby? Industrial pollutants in the air, water, or soil? Private landowners are sometimes eligible for a government cost-sharing rural clean water program.

Well water. If you have your own well, it should be located up hill and away from livestock areas. Most wells are drilled holes, lined with steel casing, 6 to 8 inches in diameter and from 50 to 400 feet or more into the ground. The well casing extends about 18 inches above the ground and has a watertight cap. The soil around the exposed steel casing should be graded to prevent water from collecting on the surface. Hand-dug wells can be as large as 6 feet in diameter but rarely are deeper than 20 feet or so. They are often lined with cylindrical concrete tiles stacked on top of each other.

Avoid dug wells lined with timbers, stones, or galvanized sheet metal.

You will need to test your well periodically for bacterial contamination and water quality. If you have a storage tank or cistern test it also. Bacterial contamination is usually tested by your county health department. Water quality testing is often provided by your state agricultural university or a private laboratory.

Community-supplied water. If your acreage's water comes from a community source, it should arrive at your property in an acceptable form. It would be a good idea, however, to test the water for purity, as contamination can occur anywhere along the way or on your property. Also, baseline mineral and pH values provide valuable reference in the event your young or pregnant horses experience problems that may be attributed to water chemistry imbalances.

Natural waters. Creeks, springs, and streams can provide a fresh supply of drinking water but they also may be a source of contaminants. (See the following chapter.) Ponds can serve as watering spots but may become stagnant.

When in doubt about the suitability of natural waters on your property, contact your county agent and have the water tested. Also, be sure the approach to and footing around watering holes are safe. During the winter, for horses turned out on pasture, you may need to break the ice several times a day, even in a flowing stream.

Water for irrigation. If you plan to irrigate your pastures or fields, you should evaluate the quality of your irrigation water for its soluble salt content, sodium content, toxic elements, and bicarbonate concentration. Excess salt makes moisture less available so that even though a field appears to have plenty of moisture, the plant roots are unable to absorb it and they experience a physiological drought. Water with a high sodium content affects the physical properties of the soil, eventually making the soil hard and compact and increasingly impervious to water penetration.

Two toxic substances sometimes occurring in water are boron and chlorine. Toxic elements can affect the pasture or hay fields immediately or after they have accumulated over a number of years. Water which is high in bicarbonate tends to result in a higher sodium hazard.

Evaluating Water Quality

No matter what the source, the water for your horses must be pure and palatable with minimal mineral matter.

Purity

There should be no harmful organisms or decomposing organic matter in your horse's water. This means that the well (and cistern if you have one) must be free from contamination and that the water vessels you use for the horses are kept scrupulously clean of old feed, algae, dead animals, and dirt.

To test your well for purity, obtain a special sterilized container provided by your county or local Department of Health and take a sample according to the instructions provided with it. The sample must be representative and should not be from a new or inactive well. A well should be thoroughly pumped before sampling.

The laboratory will test it primarily for coliform bacteria, a large category of bacteria associated with intestinal discharge of humans and domestic and wild animals. If your test results show the presence of coliform bacteria, this indicates that your well has been contaminated since its last sanitization. Some common sources of contamination are: a crack in the well house roof, walls, or floor; an improper well cap which lets run-off into the well; a bird or mouse which fell into the well when the well cap was left open.

Chlorination. If your sample shows contamination, the health department will probably send you chlorinating procedures. Since you will be without water for a whole day during chlorination, you should draw emergency water or arrange for the delivery of other water.

Generally one gallon of 5-percent household chlorine bleach should be mixed with 10 gallons of water and added to the well. Water should then be run from the well through every pipe to all water outlets, hot and cold, including all household outlets (don't forget toilets), hydrants, and barn fixtures and spigots. Let the water run until you smell chlorine at each outlet and then turn off the water. When the chlorine water is in all the pipes, turn off your pump. Let the chlorinated water stand in the pipes, receptacles, pressure tank, and pump for eight to twenty-four hours. Do not use any water during this time. Then open all receptacles and let the water flow until the odor and taste of chlorine is gone. Two days later, retest the well. Plan for a routine well check once a year.

Soil Testing Laboratory
Colorado State University
Domestic Water Analysis Report

DATE: 11/21/88   SO6166
NAME: Mary Thomas   Lab No. W3809
ADDRESS: 6 County Road
LOCATION: Larimer County — Private well — Cistern

Conductivity: 143 Micromhos/cm (E.C. x 1,000,000) pH 6.7

	Results (mg/1)	Recommended Limits (mg/1)
Calcium	16	
Magnesium	3	
Sodium	8	20
Potassium	2	
Carbonate	<1	
Bicarbonate	67	
Chloride	2	250
Sulfate	8	250
Nitrate	7.1	45
Nitrate as nitrogen	1.6	10
Total Alkalinity as CaCO3	55	400
Hardness as CaCO3	52	300
Total Dissolved Solids	113.1	500

Additional Tests: (mg/1)

Phosphorus	<0.1	Zinc	0.02	Barium	0.02
Aluminum	<0.1	Nickel	<0.01	Iron	0.04
Molybdenum	<0.01	Manganese	<0.01	Cadmium	<0.01
Copper	<0.01	Chromium	<0.01	Boron	0.01

COMMENTS: This is good quality water for domestic and livestock use.

Sample water analysis report.

If you need to chlorinate a cistern or other storage tank, determine the volume in the pressure tank, the pump, the pipes to the cistern, and the cistern itself. Check the chlorine dosage rate recommended by your health department. If you add too much chlorine to your water storage, it can kill plants, be unpalatable to animals, and may even harm your horses. If you don't add enough, it won't disinfect.

Palatability

Fresh water from clean streams or springs or water that has been just drawn from a good well is bright, has a pleasant taste, and is naturally aerated. Air promotes digestion in horses by helping digestive juices permeate feed. Poor-smelling water indicates the presence of sewage gases from decomposing organic matter. Horses have a keen sense of smell and detect tainted water easily. A bucket of water in an untidy stable or a stale trough often absorb impurities, such as carbonic acid. Horses usually refuse to drink such tainted water.

Rainwater from roofs often contains soot, dirt, and other impurities. A green or yellow color to the water often indicates fermenting or decomposing plant matter. An iridescence is often due to the presence of petroleum products. Muddiness, from clay,

EFFECTS OF EXCESS SALT IN WATER

Amount of soluble salts*	Effects on horses and crops
Less than 1000	Excellent water
1000-3000	Very satisfactory for animals; may cause temporary or mild diarrhea in animals not accustomed to it; low palatability. May have adverse effects on many crops.
3000-5000	Satisfactory but may cause temporary or mild diarrhea in animals not accustomed to it; often refused. Must be used with very careful management practices on salt-tolerant plants.
5000-7000	Reasonably safe except for pregnant or lactating animals. Not suitable for crops.
7000-10,000	High risk with pregnant, lactating, or young horses.
Over 10,000	Unsuitable for animals or crops.

*Expressed in mg/l
SOURCE: Colorado State University Service in Action Bulletin Number 4,908.

for example, poses no big problem but it is always best to try and provide clean and clear water.

Mineral Matter

The mineral content, as well as pH and levels of salts, sulfates, nitrates, and metals, will be revealed by a chemical water analysis. Hard water, usually from deep wells and some deep rock springs, has a large percentage of minerals or solids in solution, while soft water, such as distilled water, rainwater, or surface water springs, streams, and rivers, has a small percentage of minerals in solution. As you look at a chemical evaluation of your water, keep in mind that what might constitute good drinking water for you and your horse may not be the best water for washing. (See the section on softening water, page 131.)

Although it is OK to give horses drinking water that contains moderate amounts of calcium and magnesium carbonates, water with excessive levels of these compounds can have an astringent, laxative, or dehydrating effect. Some animal owners and some veterinarians feel that hard water can cause urinary calculi ("stones" in the horse's urinary tract), but currently there is no data to support the theory.

Water that contains high levels of soluble salts such as sodium chloride, calcium chloride, magnesium chloride and some of the sulfates (see chart) can also have deleterious effects.

pH

You should also have the pH of your water assessed. Absolutely pure water has a neutral pH of 7.0, and wells usually range from 6.5 to 8.0. Some metals, such as lead and zinc, are more soluble in acid water. If water with a pH of less than 5.0 (acidic) runs through lead pipes, it may result in corrosion and subsequent ingestion by your horse of excess lead, which can be toxic. Lead pipes carrying soft water can also lead to lead poisoning. Today PVC (polyvinylchloride) water pipes are used in most residential and agricultural applications. The calcium and magnesium in water with a pH greater than 8.5 (alkaline or basic) tends to precipitate out, causing a white, crusty residue or film. The calcium and magnesium are replaced by sodium, resulting in a significant amount of sodium bicarbonate present in alkaline water.

Other Impurities

Other undesirable substances in water for horses include sulfates, nitrates, and toxic substances. Water with a high sulfate level might indicate contamination by waste or septic intrusion. It can be toxic to plants and its bitter taste makes it unpalatable for horses. High concentrations can have a laxative effect.

High nitrate concentrations in water are often caused by agricultural run-off. Be sure you are preventing nitrogen loss in your pastures and fields by using proper fertilization practices. Nitrates interfere with the oxygen-carrying capacity of the blood by altering hemoglobin and so are especially bad for pregnant and young animals.

Arsenic, selenium, barium, cadmium, and mercury above recommended limits must be removed. Check with your extension specialist and your veterinarian for specific limits and suspected problems in your area.

If you have a serious water problem, you may need to invest in a reverse osmosis water purification system. Such a system removes essentially all impurities in water by filtering the water through a semi-permeable membrane filter. Such a system is expensive, however, and so are the membrane filters, which must be replaced regularly, especially with hard water. The relatively large particles of calcium and magnesium, characteristic of hard water, quickly clog up the very fine membrane filter which is really designed to filter out microscopic entities such as bacteria and nitrates. If your water is hard and contains microscopic impurities, it would be best first to remove the minerals with a water-softening device (see section on softening water) and then to send the water through the reverse osmosis system. Due to the complications and expense, however, it may be more cost-effective to drill a new well.

Softening Water

If hard water is used to bathe horses or wash blankets it makes thorough cleaning difficult and leaves behind a white residue from the precipitated calcium and magnesium. You can overcome hardness in wash water by using water-softening equipment or adding a softening compound.

Water-softening Equipment

A popular water-softening device is the common household-salt-filled water softener, which exchanges sodium ions for the

WATER HARDNESS*		
Relative hardness	**Grains****	**Mg/l or ppm**
SOFT	0-4.5	0-75
MODERATELY HARD	4.5-9	75-150
HARD	9-18	150-300
VERY HARD	over 18	over 300

*Calculated on the basis of calcium carbonate and magnesium carbonate
**1 grain = 17.1 mg/l or 1.0 ppm

calcium and magnesium ions that cause water hardness. So, although the water is made suitable for cleaning purposes, it becomes very high in sodium (salt) and is not suitable for drinking water or for watering plants. Therefore, water from such a softener would be appropriate for washing horses and blankets but not for drinking water.

Softening Compound

You can soften batches of water that are specifically going to be used for washing horses, blankets, and equipment by adding, as needed, sodium hexametaphosphate, a water softener sold under the trade name of Calgon. (Do not confuse Calgon with Calgonite. The latter is a cleaner made specifically for automatic dishwashers and contains several harsh detergents.) This removes calcium and magnesium from the water to make it purer and subsequently a better solvent. Unlike less expensive softeners such as washing soda and trisodium phosphate, sodium hexametaphosphate is tasteless, non-toxic and has a neutral pH. Washing sodas are very alkaline which is undesirable for use on hair and skin. In addition, they precipitate minerals from the water resulting in a sludge or film in the wash or rinse water. Sodium hexametaphosphate, on the other hand, holds the minerals in suspension in the water so they can not form scum.

This water softener is useful in two ways: to increase the effectiveness of soap

Locating water tubs outside of the horse's pen ensures that the tubs will stay cleaner and may prevent a playful horse from tipping over his water tub. Cutting a hole in the panel allows the horse to put his head through to drink. This tub is half of a plastic 50-gallon vanilla barrel. Note the hose bracket on the fence panel.

be more appropriate for very hard water with a hardness score of 18 grains or more per gallon. At those rates, the 4-pound box usually available in most grocery stores goes a long way.

Store the box of sodium hexametaphosphate granules in a cool, dry place and mix as needed. Water that has been chemically softened is great for cleaning chores around the barn but is not desirable for drinking water.

Manmade Watering Devices

and to act as a thorough rinse. With cold, warm, or hot water, sodium hexametaphosphate helps soap or shampoo do a better job of cleaning as it prevents the dingy, insoluble scum from forming. As a rinse for a horse's hair or a blanket, a sodium hexametaphosphate solution removes the graying dullness left by previously deposited soap residues. It has a superior ability to combine with and sequester oily and greasy substances, which prevents them from reacting with the horse's skin or becoming trapped in the fibers of a blanket.

Some especially handy uses for a sodium hexametaphosphate solution: to sponge away the outline of bridle and saddle from a horse that has just finished working; to rinse the mane or tail without shampooing; to dampen a stable rubber for use as a dust-magnet in the final stages of grooming.

How much sodium hexametaphosphate to use depends on the hardness of the water. One teaspoon per gallon of water would be adequate for naturally soft water with a hardness of 4.5 grains or less per gallon. Two tablespoons per gallon would

If you are putting in hydrants in a temperate climate, be sure they are the freeze-proof, self-draining type and located where they are not accessible to horses. Drain and roll up hoses after each use. Troughs are fine if you have a large number of horses, but for a few, the water will probably become stagnant before it is drunk. Clean troughs regularly. If troughs are to be used in the winter in temperate climates, they should be insulated and kept from freezing with a caged tank heater. Waterers made from plastic half-barrels that held nontoxic substances (such as vanilla or vinegar) hold about 20 to 25 gallons, which is suitable for a single or double horse waterer.

Buckets work well in stalls. Two-to-five-gallon rubber buckets are often preferred over plastics, which can crack. Automatic (heated, if necessary) waterers can be shared between two stalls, two paddocks, or two pastures. In any case, no more than ten horses should share a single waterer to ensure that all horses have the opportunity to drink an unlimited amount of pure, fresh water.

CHAPTER FOURTEEN

Sanitation

....

SANITATION PRACTICES MUST BE IMPLEMENTED for your horse's health, your family's health, your relations with your neighbors, and your legal obligations. It involves proper management of manure, flies and other pests, moisture, and hazardous wastes.

No matter whether your horses are pastured or stabled, they will produce generous amounts of manure and urine daily. In an enclosed barn, the added waste products of respiration of the skin and lungs help to make the environment an ideal breeding ground for bacteria.

Urine contains urea and hippuric acid, both of which break down into products containing ammonia, a volatile gas that escapes into the air. The pungent vapor can be injurious to the eyes and lungs of both horses and humans. It can also be destructive to tack by drawing the fats and oils in the leather to the surface and combining with them to make "soap."

The combination of dung and urine is a perfect medium for the proliferation of bacteria that can actually begin destructive processes on leather. Dung and urine can also break down the integrity of hoof horn. When certain fecal bacteria ferment, their secretions can dissolve the intertubular "hoof cement." Moist manure also softens, loosens, and encourages the breakdown of hoof horn cells, more destructively than water does. (See the section on moisture management, page 138.)

Wherever there is manure, there are parasite larvae. The life cycle of all horse parasites involves leaving the horse host via the manure and then reinfesting a new host. When a horse eats from manure-contaminated ground, it ingests parasite eggs. Parasite larvae can do great internal damage to a horse as they migrate through the tissues. Besides deworming horses every two months to decrease the number and viability of parasite eggs shed, the daily removal and proper management of manure is the best way to break the parasite life cycle.

Managing the Manure Pile

Manure production on even the smallest horse farm requires constant attention. A 1,000 pound horse produces approximately 50 pounds of manure per day or about ten tons per year. Add to this from 6 to 10 gallons of urine which, when soaked up by bedding, can constitute another 50 pounds daily.

The tines of a silage fork are perfectly spaced to separate equine fecal balls from sawdust bedding.

About one-fifth of the nutrients a horse eats are passed out in the manure and urine. If the manure is properly handled, about half of those excreted nutrients can be utilized by pasture or crop plants in one growing season, with the balance being used in subsequent years. Horse manure is considered one of the most valuable of farm manures because it is quite high in nitrogen and it is "hot" or capable of fermentation. A ton of horse manure will supply the equivalent of a 100-pound sack of 14-5-11 fertilizer, as well as providing valuable organic matter and trace elements. (Fertilizer numbers designate the nitrogen-phosphorus-potash content, in that order.)

Even if manure is not to be used as a fertilizer, it must be properly managed in order to control odor, remove insect breeding areas, and kill parasite eggs and larvae.

Handling Manure

There are basically three ways to handle manure, all three of which begin with daily collection. Daily collection is essential, as dung and urine create an ideal breeding ground for bacteria and internal and external parasites. In addition, the pungent vapors of manure and urine are irritating to the eyes and lungs.

Once the manure is collected, it can be hauled away, spread immediately on a pasture or arena, or stored for later distribution as compost. Some refuse collection services are specially designed to handle manure or are willing to haul it along with other trash.

Spreading manure. If manure is to be spread daily, it must be distributed thinly and harrowed to encourage rapid drying, thus eliminating favorable conditions for parasite eggs and fly larvae. It is best that such manure not be spread on land that will be grazed by horses during the current year.

Although few problems are encountered with applying fresh horse manure to established grass pastures, it should never be applied to a garden or to newly planted trees since it is likely to burn plant tissues. Horse manure should age from six to eight weeks before being added to gardens or shrubs. For best results with new plants, work in the manure at least four weeks before seeding or transplanting.

Composting. The most common method of dealing with manure is daily collection and storage for later spreading. A good understanding of composting ensures success if this method is chosen.

Composting is convenient because the manure does not have to be hauled every day and because it reduces bulk and concentrates nutrients. It encourages the manure to release its nitrogen, diminishing odor and making it more pleasant to handle. The end-product of composting is humus, the dark, uniform, finely textured, odorless product of the decomposition of organic matter that is so valuable as a soil conditioner and additive. Manure is comprised of undigested food, digestive juices, and bacteria. The bacteria make up as much as 30 percent of the mass! Because urine (usually as soaked bedding) is a liquid, it contains more dissolved and thus readily available nutrients than do feces.

Decomposition of manure begins with the formation of ammonia as urinary nitrogen decomposes. The level of fermentation depends on the degree of compaction and moisture content of the manure pile. A well-tamped but frequently turned pile makes the best environment for the aerobic bacteria of fermentation. The pile should be uniformly moist, about 50 percent. A dry pile simply dehydrates the bacteria; a soggy heap smothers the aerobic organisms.

The next stage of decomposition is the putrefying of the insoluble nitrogen in the feces. The degree of fermentation here depends on the composition of the manure. This, in turn, was determined by the quality and amount of feed given to the animals as well as by their age, health, and condition. As the manure solids putrefy, they produce more ammonia, which is food for the bacteria.

A strong ammonia smell indicates that the ratio of carbon to nitrogen in the pile is too low. This can be from feeding excess protein, or it can be because there is little bedding mixed in with the manure and urine. A carbon-to-nitrogen ratio of about 30:1 is optimum. Manure has a carbon:nitrogen ratio of 14:1; sawdust, 400:1. A pile of pure manure would decompose in a few weeks and produce great amounts of heat; a pile of sawdust would create negligible heat of fermentation and take two to three years to become humus. The typical combination of manure and bedding usually results in a good C:N ratio for composting.

The hydrated lime sometimes added directly to the manure pile speeds up the bacterial action of fermentation, but the alkalizing action burns up valuable nitrogen and loses it to the air. Hydrated lime is also used to "sweeten" stall floors by lowering the acidity of the urine. It also causes dirt particles to clump, allowing air to get at and penetrate the wet soil more easily, and thereby dry the floor. The negligible amount of lime added to stall floors will not affect a ma-

nure pile significantly one way or another.

The final phases of composting are the death of the bacteria and the breakdown of fibers. As the bacteria die and decompose, they release their stored nitrogen. As the fiber breaks down, carbon dioxide and water are released, decreasing the bulk of the manure by up to one-half.

The process of decomposition of a manure pile can take anywhere from two weeks to three months or more and the quality of the resulting product will vary. Managing a pile properly will kill the parasite eggs and larvae, prevent flies from breeding, and result in a good-quality fertilizer.

ABOVE:
After cleaning a stall or pen, sprinkle the wet spots with hydrated lime.

BELOW:
Take the time to form a tidy manure pile.

To this end, it is best to have three manure piles: one ready to spread, one in the process of decomposing, and one to which fresh manure is being added daily.

Before starting a pile, it is best to check local zoning ordinances. Be sure the pile is out of sight and smell of residences and downwind from the stable and the house. All of this, and the pile must also be convenient for daily dumping and periodic hauling.

If possible, the piles should be located on a sloped concrete floor with 4-foot walls. The piles can be covered with a roof, plastic sheeting, or earth. An open pile is subject to drying by the sun and leaching of nutrients by rain and melting snow. If an open pile must be used, it should be about 6 feet high and 6 feet wide and can be added to in length as needed until hauling is convenient.

Flies and Other Pests

Flies

Stable flies, horseflies, deerflies, horn flies, and face flies are all blood suckers and create problems for horses and their owners. Because stable flies are usually the most common pest, I have used them as the primary example here.

Stable flies are the same size as common house flies. Both male and females draw blood, commonly feeding on the lower legs, flanks, belly, under the jaw, and at the junction of the neck and the chest. When they have finished feeding, they seek shelter to rest and digest. Their bite is painful; some horses have such a low fly-tolerance threshold that they can be driven into a frenzy or panic into an injurious run. Even rather tough horses may spend the entire day stomping alternate legs, causing damaging concussion to legs, joints, and hooves and resulting in loose shoes and loss of weight and condition.

Stable flies breed in decaying organic matter, and moist manure is a perfect medium. A female often lays twenty batches of eggs during her thirty-day life span, each batch containing between forty and eighty eggs. The eggs hatch in twenty-one to twenty-five days. When the eggs hatch, the adult flies emerge ready to breed. (If you have seen small flies and thought they were immature stable flies, you were probably looking at a different type of fly.) The number of flies produced by one pair of adults and their offspring in thirty days is a staggering figure, in the millions.

That is why fly prevention is the best way of keeping the fly population under control. It is based on removing breeding grounds, controlling moisture, and using insecticides and other fly control measures.

Removing breeding grounds. Manure management and moisture control are the two key ways to discourage flies and encourage the health of your horses. Remove manure and wasted feed daily from stalls and pens and either spread it thinly to dry or compost it (see "Managing the Manure Pile," page 133) to take away the breeding grounds.

Keep moist areas to an absolute minimum. Be sure there is proper drainage in all facilities. Repair leaking faucets and waterers. Eliminate wet spots in stalls and pens by clearing bedding away, liming, and letting the ground dry out. (See more on liming later in this chapter.)

Fly predators. An additional way to break the life cycle of flies is to use fly predators. This method of biological control is safe and nontoxic and, if properly implemented, requires much less labor for a greater degree of control than many insecticide-based methods. These tiny nocturnal stingless wasps lay eggs in the pupae of the common housefly, the biting stable fly, the horn fly, the lesser house fly, the garbage fly, and the blow fly. The wasp eggs utilize the contents of the pupae as food, thereby killing it before the fly can even develop. The wasps stay within

200 feet of where they hatched and work while you sleep. They are harmless to animals and people. Any supplemental methods of control involving insecticides must be carefully implemented or they will wipe out the predator population along with the flies.

Chemical controls. Much less will have to be done with insecticides and other control measures if manure and moisture are handled properly. The indiscriminate use of any form of insecticide (a chemical that kills flies) or repellent (a chemical that keeps flies away) can result in the development of resistant strains of flies as well as harm to horses, humans, and the environment.

There are many forms of insecticides and repellents. Long-term (up to six weeks) *residual insecticides* are designed to be applied on fly resting sites, such as on rafters or in bushes. *Fogs* and *mists* are to be used daily, either expelled in the barn air using an automatic timer or applied to the horse's body with a hand-held mister. *Impregnated strips* are useful for enclosed areas such as tack rooms, feed rooms, and offices.

Baits, including sticky paper, sweet fluids, or sex attractants, can be used in areas of heavy accumulation but require emptying or cleaning frequently. Flies can also be eliminated by black light electric fly attracters.

In addition, there are *chemical larvicides*, which can be applied to manure piles or fed to horses. Those fed to horses pass through the system undigested and then begin their work on the larvae. The distribution is unequal in the manure, however, and the larvicides have an active life of only about a day.

Topical spray repellents seem to be the method of choice, most of them containing some amount of insecticide as well. They can also contain sunscreens, coat conditioners, and products that keep the repellent on the hair shafts longer. Almost all repellents attract dust and dirt. Be certain that you read all labels carefully and know whether a product is to be used full strength or diluted.

In addition to spray-on, wipe-on, and stick repellents, there are *impregnated strips and tags* that can be attached to halters. These are especially helpful in controlling face flies, which have sponging mouth parts and feed on mucus around the eyes and nostrils. Some degree of relief can also be afforded the horse by using *fly shakers* attached to the crown piece of a halter or the browband of a bridle. These strips mechanically jiggle the flies off a horse's face when it shakes its head. Mesh *fly masks* prevent face flies from landing around the eyes. Cool, open-weave *fly sheets* keep flies from pestering the horse on its body.

One of the best means of rodent control in and around your farm buildings is cats.

Rodents

Rodents can also cause damage and health problems if they go unchecked. They can carry bubonic plague and rabies; they can damage tack and make a mess in a feed room. All feed must be stored in rodent-proof containers. Although poisons and baits can be used to control rodents, good sanitation and cleanliness, plus a few cats, work best! If a mouse or two sneaks into a tightly enclosed area (such as a tack room) where cats are not allowed free access, simple traps baited with a dab of peanut butter will eliminate the population before breeding begins.

Moisture Management

An old wives' tale suggests that if you have a horse with poor-quality or dry hooves you should let the water trough run over, to force the horse to stand in the mud. An alternative approach is to slather hoof dressing on three times a day. While the basic intention of these recommendations is good, in many cases they are absolutely the worst things you could do to your horse's hooves. As I have mentioned, moisture can be a breeding grounds for flies and mosquitoes and provides good conditions to allow parasite larvae to remain alive.

What the recommendations do not address is the fact that some hooves that appear cracked and chipped have become that way from too much moisture, rather than not enough. First you should become familiar with the moisture balance in a normal hoof; then you can tailor-make decisions for your horse according to your facilities and management situation.

Because the modern riding horse evolved on semi-arid plains, the healthy hoof is designed to be dry and hard. The hoof wall is a part of a spring mechanism that encloses the inner structure of the foot. The moisture balance in the wall controls the strength of the spring. In a normal hoof, the outer layer is dense and tough with a moisture content of 15 to 20 percent, while the inner layer averages about 45 percent. The two layers are joined together by interlocking sheaves of insensitive laminae (from the outer layer) and blood-rich sensitive laminae (from the inner layer).

Adequate moisture in the inner layers is provided by blood and lymph vessels in the laminae. Moisture diffuses outward from the moist sensitive laminae toward the dry, hard outer wall. When blood circulates freely to and from the hoof, the dynamic balance of moisture is operating at an optimal level. When there is a lack of exercise, from inactivity or from a disease of the hoof such as laminitis or navicular, circulation is interrupted to the point that moisture is no longer provided through the blood flow.

The spring-like action of the hoof is only as good as the moisture balance between the two hoof layers. The moist inner wall expands and pushes against the dry hard outer wall, which resists in turn with a curling, contracting force. This dynamic balance allows the healthy hoof to absorb shock as it "deforms" when the weight of the horse descends on it (the hoof landing on the ground) and then reforms to its normal shape when the force of the weight has lessened (the hoof in the air or at rest on the round).

A graphic demonstration of this dynamic moisture balance can be observed when your farrier nips one complete round of wall from your horse's hoof. If you can beat your dog to the trimming, put it in the sun, and in no time at all, the once hoof-shaped paring will curl into a small circle. Because the inner layer has lost its moisture and its force of expansion, it can no longer oppose the spring-like contraction of the dry outer layer.

If your horse is getting adequate exercise, the moisture level of the inner layer remains constant, so you don't have to worry about your horse's hooves curling up into small circles! If you have ever seen a horse with contracted heels, however, you probably have a pretty good idea of how and why they got that way. The horse may have stood in a stall, quite inactive, and the moisture was not delivered through the blood to the inner hoof layer. The inner moisture was diffused out through the dry outer wall more rapidly than it was replaced. Therefore, the spring-like outer hoof layer won the contest of strength and the hoof curled inward, much as the trimming did in the sun.

In contrast, a hoof that is kept too soft, as when the horse is continually standing in water or repeatedly daubed with too much hoof dressing, contains too much moisture in the outer layer. The soft outer layer doesn't stand a chance of opposing the pressure from the inner layer, and the hoof spreads out like a pancake. Excess moisture weakens the integrity of the layers of hoof horn, resulting in soft, punky hoof walls that peel and separate into layers.

Usually the inside quarter goes first, then the outside quarter, and finally the heels. When the heels separate into two layers, the inner layer collapses inward toward the sole and the outer layer squishes out over the edge of the shoe, leaving no hoof wall at all to bear the horse's weight!

Too much moisture also makes a horse's soles soft and susceptible to sole bruises and abscesses. The continuous use of full pads can similarly disrupt the natural moisture balance of the hoof. Pads seal hoof packing against the sole creating a moisture-rich environment, perfect for thrush and fungus infections. The sole cannot respire normally. In addition, using pads can make the sole tender and the horse pad dependent. Just watch a horse that has been wearing pads be led barefoot and you will likely see him say "ouch" with every step. This can usually be corrected, however, by competent shoeing that leaves the sole exposed to get dry and tough.

The effects of repeated wet/dry conditions raise havoc with the hoof's attempt to maintain the moisture status quo. Research has found that the condition of hooves worsen during hot, humid weather, especially where horses are turned out at night (see chapter 16). Typically, horses walk around in dew-laden pastures all night and then are either left out where the sun will dry the hooves or put in a stall where the bedding dries the hooves. Horses that receive daily baths or rinses or those that repeatedly walk through mud and then stand in the sun experience a similar decay in hoof quality. In both cases, the hoof is undergoing a stressful moisture expansion/contraction that is damaging to its structures.

If you want a first-hand example of how drying such situations can be, stick your fingernails in some fresh mud and let it dry.

Your hooves will probably begin to split and crack after just one episode! Mud has the effect of drawing out moisture and oils and tightening pores — as in a poultice or a mud facial.

In the process of drying out, the outer layer of the hoof wall will attempt to bend or warp but cannot really do so because of the hold the inner layer has on it. So instead it will develop cracks and checks to relieve the stresses from the shearing forces of the opposing layers. The cracks are then packed with more mud and dirt so that they can't close and continue getting larger and spreading upward.

In addition, the drying process breaks down the cellular cement that holds the hoof horn together. The horn cells disintegrate, leaving a frayed, pulpy mass of dry fibrous tubules. Drying mud and water also leach out essential nutrients and oils that are responsible for maintaining hoof flexibility.

Soft or brittle hooves do not hold nails very well, so it is difficult for your farrier to find a solid piece of horn to nail to; consequently the shoes become loose in a much shorter period of time than normal. Commonly such a horse loses shoes, unless they are reset every three or four weeks. Clips may help hold the soft hoof together and take some of the stress off the nails, but it is important to get at the cause of the poor-quality hoof horn.

Hoof dressings. Because excess moisture can be so damaging to a hoof, some researchers recommend never applying greases or oils to the hoof. Because of the two natural coatings on the hoof, dressings may not be able to penetrate anyway! A waxy covering, the periople, located at the coronary band, is visible as an inch-wide strip encircling the top of the hoof. The protective, varnish-like outer layer of the hoof wall, the stratum tectorium, is composed of very hard cells. Both coatings retard moisture movement from either direction—from the outside environment into the hoof or from the inner

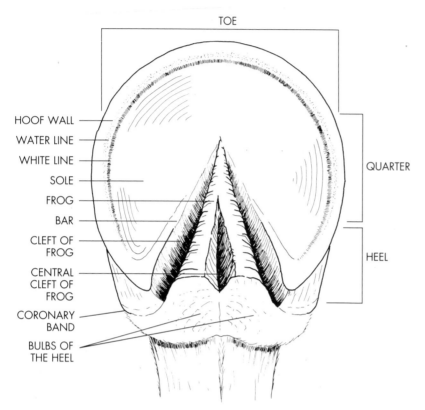

TOE

HOOF WALL
WATER LINE
WHITE LINE
SOLE
FROG
BAR
CLEFT OF
FROG
CENTRAL
CLEFT OF
FROG
CORONARY
BAND
BULBS OF
THE HEEL

QUARTER

HEEL

**Drawing #29
Parts of the hoof.**

layer of the hoof to the outside.

About the only time hoof dressing is warranted is when the bulbs of the heels have become so dry that they are beginning to crack. In order to restore their pliability, rub a product containing animal grease, such as lanolin or fish oil, into them daily until the desired result has been achieved. Petroleum-based products are fine for lubricating steel but are counterproductive when used to condition hooves or leather. Petrolatum is thought to emulsify the hoof's natural oils and actually remove moisture.

Problem hooves can be attributed to genetics, improper nutrition, or inadequate management. Nevertheless, horses that have inherited the potential for good hooves and are fed a balanced ration can still develop poor hooves with bad management. A horse with low-quality hoof horn faces a long road to recovery. It takes up to a year for new growth from the coronary band to reach the ground. In the meantime, your farrier can try to minimize trauma to the weak hooves

by using fewer nails, thinner nails, clips, special hand forged shoes, plastics and acrylics, or even glue-on shoes.

A horse owner, however, has the greatest control over hoof condition, since management practices can greatly improve or worsen hoof quality. Prevention of problem hooves is worth the time and effort invested. One of the best guarantees that your horse's hooves are operating at an optimum moisture exchange is to assure that the horse is getting regular exercise. So, saddle up and ride to better hoof health!

Thrush. Another common problem with horses that stand in dirty, wet footing is the condition called thrush. Thrush is a decomposition of hoof tissues that usually starts in the clefts of the frog. Anaerobic bacteria, those that survive and flourish without air and are present almost everywhere on the farm, are sealed in the clefts by mud and manure and kept moist by urine and muddy conditions. The bacteria destroy the hoof tissue and produce a foul-smelling, black residue that you will never forget once you have smelled it. Erosion in the clefts of the frog can become so extreme as to cause lameness and reach sensitive tissue.

Prevention requires frequent inspection and cleaning of the hooves. Treatment for thrush begins with your farrier paring away the diseased tissue and flushing the area with hydrogen peroxide, a commercial thrush medication, or a very dilute chlorine bleach solution. You will then have to keep your horse's hooves scrupulously clean and dry. Remember, thrush bacteria cannot grow in the presence of air.

Hazardous Wastes

Guarding your horse against chemical poisoning is essential because horses like to investigate unknown things by nibbling and tasting.

All toxic substances must be stored in tight, well-labeled containers. This includes rodent poison, insecticides, herbicides, and lead-based paints. Lead-based paints should never be used on horse buildings or fences. Be sure to read all labels thoroughly and follow directions carefully. Any unlabeled substance should be discarded in a safe manner.

Chemical poisoning can often occur unknowingly. If you think that, in general, "since a little is good, more is better," you may seriously harm your horse. Never overdose horses on antibiotics, other drugs, or nutritional supplements. Don't feed treated grain or seeds that were meant for planting or you may be giving your horse a dose of mercury! Although treated grains often have a pink or reddish hue, sometimes they look just like feed grain.

Don't give horses feed that was meant for cattle, sheep, or goats. Often these ruminant feeds contain urea, a source of non-protein nitrogen designed for ruminants. And some cattle feeds may contain growth stimulants that can be permanently damaging to the nervous system of the horse.

Don't let horses near junk or vehicles. Using lips and teeth to inspect things they may ingest toxic paints, antifreeze, or battery fluids. Protect horses from all fumes: vehicles, paints, and solvents. Don't apply insecticides or herbicides near feed or water areas and be aware of wind drift when you are spraying.

Prevent organophosphate poisoning. Some dewormers and some insecticides contain chemicals belonging to the organophosphate group. These chemicals have a cumulative toxic effect in the horse and must not be given within a certain number of days of each other.

Additional Disease Prevention Principles

Disease and infection are usually spread either by physical contact, by contaminated feed or water, or by air-borne antigens. If you have disease or infection on your acreage, you must work closely with a veterinarian to bring it under control. A combination of treatment, disinfection, and quarantine will eliminate the spread of disease or infection and eradicate the disease-causing organisms.

Sunlight, especially dry, hot air is a powerful disinfectant. Specific chemicals are effective against certain organisms. Your veterinarian will advise you what to use.

Besides the healthy practices outlined in this book, problems can be minimized by routine immunization and quarantine. All horses should receive a yearly booster each spring for sleeping sickness, (equine encephalomyelitis) and tetanus. In addition, two immunizations per year may be necessary for the respiratory diseases influenza and rhinopneumonitis. It may also be advisable for you to vaccinate your horses against strangles and rabies.

All new animals should be quarantined upon arrival and observed for at least a week before mixing in with resident horses. Any horses who leave the farm temporarily, and especially if they have been exposed to a large number of other horses, should be quarantined upon return.

CHAPTER FIFTEEN

Fire Prevention

A FIRE CAN BE ONE OF THE MOST DEVASTATING OF life's experiences — physically, financially, and emotionally. Knowledge, preparation, and forethought, however, can prevent barn fires.

Fires need a source of heat, oxygen, and fuel. While heat can be simply sunlight or friction, more commonly it is electrical failure from rodent-damaged wires or from improperly managed portable appliances. Fires can also be started by lightning, the open flame of a match or cigarette, or from the spontaneous combustion of hay or bedding.

Spontaneous combustion. For the first two or three weeks after the hay is cut, bacterial action creates heat, especially in alfalfa and clover. If the temperature of a hay stack is more than 150 degrees Fahrenheit, it should be checked frequently for an increase. If it reaches 175 degrees, it will likely begin charring. At 185 degrees, it should be moved out of the barn, with firemen present. Often a stack will smolder until it reaches the oxygen at the outer edges of the stack, and then it will explode or burst into flames. Damp bedding or grain can similarly com-

bust spontaneously.

Most horse barns are well aerated, providing a good source of oxygen to spur on a fire. Fuels for a fire can be gases, liquids, or solids. Propane, gasoline, alcohol, liniments, paints, hoof dressings, hay, bedding, grain, tack, and combustible building components can all add fuel to a fire.

Safety Practices

You can see, therefore, why it is essential to institute fire-preventative management practices.

● No smoking in any buildings. Put up signs and enforce the rule with no exceptions. Provide sand-filled containers for guests to dispose of their smoking materials.

● Keep hay and bedding separate from the stable and tack room, with a 100-foot buffer zone. Be sure all hay is well cured.

● All appliances should be disconnected when not in use or supervised and routinely inspected. This includes radios, clippers, water heaters, pipe heating tape, treadmills, and bug electrocuters. Stall lights should be in cages or heavy glass covers. The improper use of heating units — electric, kerosene, or propane — and infrared lights are among the biggest causes of barn fires today.

● Be sure wiring is in conduit so rodents (or horses) can not chew it. Don't overload any plugs or circuits.

● Use fireproof materials wherever possible. Although they may be more costly, they can often result in lower insurance rates. Fire walls or sliding fireproof doors between sections of the barn will minimize the spread of fire. They can be located between storage and stable areas and can be used to divide the barn itself into smaller sections. Consider using concrete or block walls or fire-retardant wood with paint, sealer, or coating.

● Create an airtight, fire-rententive environment. Use steel and masonry as much as possible in the construction of the barn. If using wood, choose a fire-retardant type that is pressure-impregnated with mineral salts to reduce the flammable elements of wood. The roof should be of unburnable metal or a treated wood. Fiberglass panels, in contrast to glass windows, have a higher resistance to heat and will not explode like glass.

● Locate fire extinguishers of the right type in the right locations. Check them regularly and know how to use them. The National Fire Protection Association classifies fires under one of several categories: Class A fires involve ordinary combustible materials such as wood, paper, cloth, rubber, and many plastics. Class B fires are fires of flammable liquids, oils, greases, tars, oil base paints, lacquers, and flammable gases. Class C fires involve energized electrical equipment where the electrical nonconductivity of the extinguishing media is of importance. Note that if you use water to extinguish an electrical fire, the water stream may conduct electricity and you may receive a shock.

● Keep your barn clean of dust, cobwebs,

chaff. Keep debris from accumulating in and around the barn. This includes loose hay, manure, lumber, oily rags, twine, etc. Manure should be removed daily to a site away from the barn. Grass around the barn should be mowed regularly. Locate any gas pumps or storage tanks at least 25 to 50 feet from the barn. Park machinery and vehicles at least 12 feet from the hay and stable. Combustible fluids, such as insecticides, clipper wash, pesticides, and veterinary supplies, should be stored in tight containers and in small quantities. Large quantities should be stored away from the barn.

● Lightning rods, antennas, and wire fences that are attached to the barn must be grounded. A lightning rod should have a Underwriters Laboratory (UL), Inc., label ensuring that the rods and the grounding element have been made to specification.

● Consider installing a fire detection system. Household smoke detectors don't work well in barns, as dust makes them inoperable. Choose one that is specially designed for barns and is easy to recognize and hear. You can have it sound in the barn, the house, and/or the fire department.

● In addition, you may wish to consider installing a fire-fighting system, such as sprinklers. Note that water sprinklers require a good source of water. Automatic chemical spraying systems are self-contained and work much as the dry-chemical hand-held extinguishers do.

Emergency Procedures

● Design an emergency fire plan. Have a fire drill regularly so that all family members understand the priorities and dangers. In general, the goals are to protect humans, horses, equipment, and buildings, in that order.

● If possible, promptly put out small fires with extinguishers or by smothering them with blankets.

IN CASE OF FIRE

Immediately notify your local fire department at: _____
(TELEPHONE NUMBER)

Say: "I have a stable fire at"_____
(SUPPLY YOUR STABLE ADDRESS)

If necessary, give brief directions to your location: _____

_____

_____

Evacuate humans and horses from the area as quickly as possible. If manpower is available, fight the fire.

HAZARDS

- electrical appliances including heat lamps, space heaters, and fans
- flammable liquids such as kerosene, gasoline, paint, pesticides, cleaning agents, and fertilizers
- excessively long grass surrounding the barn
- trash and manure stacked against the barn
- dirty, dusty light fixtures, fuse boxes, and switches
- hay stored in the barn
- poor ventilation in the hayloft
- tractors and other machinery stored in the barn
- any clutter that would make your property inaccessible to fire trucks

SAFEGUARDS

- test fire detectors monthly
- test fire extinguishers every six months
- recharge fire extinguishers yearly
- stage regular fire drills
- diagram your barn so that in case of an emergency you or any other person in the vicinity will be able to act as quickly and efficiently as possible. On the diagram, plot your barn's floor plan. Include the location of exits, fire alarm, fire extinguisher(s), first-aid equipment and equine inhabitants. Also keep an up-to-date count of the number of horses in your care along with a list of their owners' names, addresses and telephone numbers.

- In a larger fire, account for all humans first. Then your actions will depend on the number of people you have available.

- Notify the appropriate authorities. Keep the fire department number and your address and location and directions by the phone. It also helps to have a prepared statement at hand that you can read in the event your thoughts are scattered. "I have a stable fire at . . . ."

- Have halters and lead ropes by each stall. Evacuate horses in a planned manner, never risking a human life for a horse. Lead horses out, possibly using a blindfold. Don't spend a lot of time on this, however, as it often does not make a difference. Never drive a horse out of a barn. If a horse refuses to leave, walk away from the horse — do not jeopardize your life.

- Put horses in an enclosed area or tie them to a safe strong post so that they cannot return to the barn or tear around endangering fire fighters.

- Keep all entrances clear, as well as the areas surrounding the fire.

- Once all horses and humans have been

removed, reduce the oxygen to the fire by shutting all the windows and doors. Move all equipment and machinery. Protect near-by structures by hosing them down and soaking the base area of the building as well.

● During all of these procedures, be sure all children are kept away. Without fully understanding the dangers of a fire, a child may reenter a barn to save a pony or horse (which may have already been removed from the barn) and be overcome by fumes or smoke. Most fire fatalities are caused by fumes and smoke, not by fire. Resist the temptation of going back into any part of the barn to get your favorite trophy or saddle.

🐎 CHAPTER SIXTEEN

Daily Routines and Management Plans

....

HORSES ARE CREATURES OF HABIT AND ARE VERY content when good management practices are implemented on a regular basis. Establishing daily, weekly, monthly, and yearly routines will increase your horses' well-being and minimize your veterinary bills. These routines can also bring a sense of contentment and order to your own life.

The daily check. You should establish a routine where in a couple of moments you can assess the overall health and well-being of your horse. First you must have a good sense of what is normal for each horse. Like humans, each horse is very different. Your examination should begin with a good visual inspection, first noting the overall stance and attitude of your horse. Is he alert yet content? Or is he lethargic, inattentive, or anxious? Next pay close attention to the legs. Look for any wounds, but also for swelling or puffiness. Examination of the legs is more complete when accompanied by palpation. This requires that you develop a feel for what is normal texture and temperature for your horse's legs.

Note whether your horse has finished all of his feed from the previous feeding and if he has taken in a sufficient amount of water. Glance at his manure to see if it is of normal consistency. It should be well formed, yet the fecal balls should be almost "brittle" if you try to break one in half. Hard, dry balls indicate the horse is not taking in enough water. Loose sloppy piles indicate that the feed is inappropriate for the horse (either it is too rich, or the horse is taking in too much salt and water) or something (such as an antibiotic) has altered the microbial population, resulting in a loose stool. Excess slime or mucus indicates an irritated gut. If there is whole grain in the feces or long pieces of fiber from hay, it means the horse is either gobbling his feed without chewing, that the rate of passage of feed is too quick to allow digestion, or that the horse can not chew his feed thoroughly and needs dental care. The presence of parasites in the feces indicates a long overdue need for deworming.

Look at his stall or pen and his body for signs of rubbing, rolling, or pawing. If you have reason to suspect there might be any problem, you may wish to take your horse's temperature, pulse, and respiration as well as doing a pinch test for dehydration and capillary refill time to assess his circulation.

ABOVE:
Design a way to accumulate baling twine to prevent accidentally feeding it to your horses or letting it lie around to cause accidents.

ABOVE RIGHT:
Quilted blanket.

These are things you can learn from your veterinarian. (See Glossary.)

Daily Routine

In the morning, feed your horse his hay and give him a quick check. Feeding him hay first takes the edge off his appetite and decreases the chance that he will gulp his grain. Be sure he has enough clean water. Feed the grain ration.

Sometime after the horse has finished his morning ration, take him out of the stall or pen and tie him in a grooming area. As you pick out his hooves and groom him, give him a closer check. Then either turn him out for his daily exercise or ride him.

Clean the stall or pen as described in the following section and then re-bed it. Return the horse to his stall or pen for the evening feeding. Again, feed hay first, check the water, and then feed grain.

Cleaning a Stall

It is far easier to clean a stall when the horse is elsewhere. Remove the dung piles, using a special fork with close-set tines (a silage fork works the absolute best). Then search for the spots of wet bedding and remove them. Expose the stall floor to dry by banking the bedding against the stall walls. Lime the wet spots, if desired. (See chapter 14 on sanitation.) Let the stall floor dry all day, with barn doors and windows open if possible.

In the evening, rake the dry "old" bedding back in the area of usual defecation and urination. Add fresh bedding, if needed, to the area where the horse lies or stands. Every week or two, depending on how heavily the stall is used, remove all the bedding and start fresh. Before you return your horse to his stall, pick out his hooves and give him a good brushing.

Stable Clothing

If you will be keeping your horse in a stall a good deal of the time, you may need to invest in some stable clothing for him. Since blankets replace the natural coat, they will discourage a horse from growing a winter coat and cause him to shed it if he already has one. Too light a blanket simply won't make much difference, but too heavy a blanket is unhealthy as it may make the horse sweat and then chill.

You may find that you will need to use an additional surcingle or body roller to keep heavy blankets from rotating on the

TYPES OF BLANKETS

SHEET. Made of linen or cotton, so will shrink — buy about 4 inches larger. Light, sturdy protection from dust, flies, and drafts. Used during spring or fall or as liner in winter.

COOLER. Usually made of wool, so hand-wash cold. Usually unfitted and draped loosely over horse after exercise while cooling out. Often held in place with straps at browband and tail.

FLY SHEET. A fine mesh with large holes. Horse does not get hot; helps to keep flies off.

WINTER BLANKET. Many types: cotton duck with wool lining; "Congress" quilted nylon with polyester fiberfill inside and perhaps artificial fleece lining. Usually need very large machine to wash; otherwise dry clean if label specifies. Will have two belly straps, chest strap, and perhaps hind leg straps and tail strap. Hoods are optional.

TURN-OUT RUG OR NEW ZEALAND RUG. Water-repellent heavy canvas with wool lining. For clipped horses or those with "summer" coats that are to be turned out in the winter.

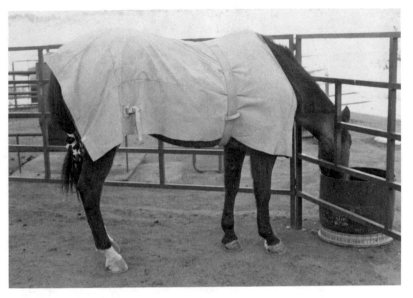

Canvas New Zealand turn-out blanket.

horse's body. The projections on the top of a body roller (something like the handholds on a vaulting surcingle) discourage a horse from rolling over on his back and getting "cast" in his stall. If a horse rolls completely over near a stall wall, he may get his legs trapped in a curled-up position between his body and the stall wall, preventing him from getting up. It can be dangerous for a horse's digestive system if he is trapped on his side for longer than an hour or so; therefore, it is a good idea to keep an eye on a horse that is prone to roll in his stall. Never put a blanket on a sweaty or dirty animal and leave him unattended.

Proper blanket fit is paramount. Blankets that are too small can cause rub marks and sore spots on the withers, shoulder, chest, and hips. Extra-large blankets have the reputation of slipping and twisting, possibly upside-down, which can cause the horse to become dangerously tangled. Measure your horse from the center of his chest to the crease between his hind legs to determine blanket size. See chart on the next page.

Blanket linings must be of a smooth material, such as nylon or silk, to prevent damage to hair, especially the mane area nearest the withers.

Blankets must be kept clean or they will cause discomfort and possibly disease to your horse. Manure and mud on blankets fatigue the material and cause them to tear and disintegrate. Have any damaged blankets repaired and keep all leather straps and fittings well oiled.

Overheating. Overheating can be a real problem with blanketed horses. Horses are often turned out to exercise in the same blanket that they wore all night. What is appropriate for low night-time temperatures in a barn is not necessarily desirable for a sunny paddock, even though there may still be snow

If you winter-clip your horse, you will have to provide extra shelter and clothing until spring.

AVERAGE BLANKET SIZES*

Hands	Size in inches	Category of horse
10	48-54	Small pony
11	56	Pony
12	60	Pony
13	64	Pony
14	66	Yearling
14 H	70	Two-year-old
15	72	Small horse
15 H	74	Thoroughbred, Saddlebred
16	76	Thoroughbred, Saddlebred, Quarter horse
16 H	78-80	Warmblood, Draft cross, Quarter horse
17	82-84	Warmblood, Draft cross

Measured in inches from the center of the chest to the center of the tail.

on a cold, still, sunny day.

Waterproof blankets do not allow heat and moisture from normal body respiration to escape. Too many layers can cause the horse to sweat, then chill, which lowers the horse's resistance by sapping his energy. This is an open invitation for respiratory infections. Check for overheating by slipping a hand under the blanket at the heart girth area.

The clipped horse. Horses that have been body-clipped or trace-clipped must be blanketed. Clipping allows a horse to be more easily worked, cooled out, and groomed in the winter months. The first clip may occur in October and may need to be repeated several times throughout the winter and early spring, depending on the horse's work, blanketing, and housing. If you want your horse's head and neck to keep a short coat as well, you may wish to consider a hood made out of the same material as your blanket. Be sure the eye holes and ear holes fit your horse properly or it may cause him to rub. Hoods can be modified by your tack repair shop. To prevent damage to the mane and rub marks on the face, be sure a hood is lined with a smooth material such as nylon.

on the ground. An unblanketed dark horse has the capacity to absorb much of the sun's energy and can actually feel hot to the touch

Wraps. In addition to blankets, you may wish to use leg wraps for your stalled horse. They are not necessary unless the horse is in heavy work or is accustomed to wearing bandages. *Stable wraps* (or standing bandages), which usually extend from below the knee or hock down to the fetlock, can be made of wool, flannel, or fleece, with or without a cotton quilt underneath. They offer warmth and protection and keep a horse in heavy exercise from "stocking up" (accumulating fluids in the legs) as he stands in his stall.

Stable wraps are not like *exercise wraps*. Since the latter are designed for temporary support, they are applied with a greater amount of tension. Stable wraps are often left on overnight (but should not be on more than twelve hours), so they must not interfere with circulation in any way. Stable wraps are applied much like shipping wraps, in thick layers with moderate tension.

Attending to a Horse on Pasture

If your daily chores involve pasture horses, you will have different considerations. Just because pasture horses might be farther "out of sight," do not let them be "out of mind." Take the time to see them every day and give each one a visual inspection.

Teach your pasture horses to respond to a call or whistle by feeding them when they come. To discourage fighting, place the feed in more piles or tubs than there are horses, and locate the feed far apart.

Hay Carrier

If you are looking for a way to carry a few flakes of hay out to the pasture, you may wish to make a simple hay carrier. It will keep you from carrying the hay in front of you and having the hay fall down the front

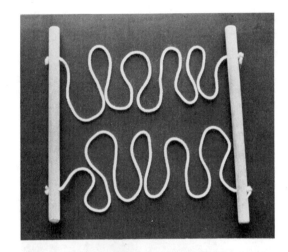

ABOVE AND BELOW
Hay Carrier.

of your shirt! You can load up rations for about four horses and carry it much like a hay suitcase.

To make one, first locate an old broom or mop handle about ¾ inch in diameter, and cut off two 16-inch pieces. Round the ends to eliminate splinters and snags. Drill a ¼-inch hole about 2½ inches from both ends of each stick. These will be the handles.

Cut two pieces of ³⁄₁₆-inch cord (nylon preferred), each about 5 feet, 4 inches long. Thread one end of one of the cords through the top hole in one of the handles and tie an overhand knot in the end of the cord so it won't slip back through the hole. If you are using nylon or poly cord, hold a lighted match to the knot for a few seconds to seal it. Do the same with the other end of the cord and the top hole in the other handle. Then repeat for the second cord and the bottom holes in each handle. When spread out, you have a rectangle. The hay is placed on the ropes, the handles are brought over the top of the hay and it's ready to carry.

Use safe feeders.

Pasture Grain Feeder

A homemade bucket-style tire feeder works well to feed grain or hay cubes in fairly level pastures. To make one you need:

● two tires — usually 14-inch or 15-inch tires work the best

● one bucket — a rubber or Fortex-type bucket works well, and you can use one with a leak or a badly damaged handle

● no. 11 galvanized wire — this is a smaller diameter-wire than no. 9 fencing wire

● two pairs of pliers or Vise-Grips

● chalk

Stack up the tires, one on top of the other. Mark the bottom rim of the top tire and the top rim of the bottom tire in four places and label the pairs one through four.

Using an electric drill, make a ⅜-inch hole at each mark about one inch inside the edge of the rim in the valley formed by the bead of the tire.

After drilling all eight holes, stack the tires in their original position, so that the holes in the bottom side of the top tire line up with the holes in the top side of the bottom tire. Cut four 18-to-24-inch pieces of wire. Thread one piece of wire through both number-one holes twice.

Grab each end of the wire with a pair of pliers and pull the tires tightly together. Twist the wires together for one to two inches, then cut off the excess. Bend the twisted tail into one of the tires. Then do the same with the pair of holes numbered three, then two, and finally four.

Remove the bail (handle) from the bucket, and set the bucket in the tires. On the top tire, mark the location of the bucket's handle eyes. Drill two holes in the tire. Cut two 12-to-16-inch pieces of wire. Loop one piece of wire through both the hole and the eye twice. With the pliers, tighten the wire and twist it together for one or two inches. Cut off the excess. Fold the twisted end securely and tuck it under the tire rim.

Pasture Manners

Even if you are not regularly riding a pasture horse, you should still take the time to ensure that he retains good manners so that you can catch and handle him for routine farrier and veterinary matters and turn him loose safely.

Catching and turning loose. Nothing can be so frustrating as a horse that refuses to be caught, and few things can be as dangerous as turning a horse loose that jerks away prematurely, often wheeling and kicking as it gallops off. The horse that keeps just out of catching distance from you, and the horse that pulls away when being turned out, have one thing in common — disrespect of both humans and equipment. It is best to prevent such habits from developing by holding regular sessions for all horses. Leaving a halter on any horse, especially one on pasture, is dangerous because the halter can become hooked on a fence, tree, or even the horse's own shoe if he is scratching. Leaving a halter on a horse full time also side-steps the important lesson of haltering and unhaltering.

First, all equipment must be well-fitted and strong. A loose halter is ineffective and equipment that breaks usually rewards the misbehaving horse with freedom. Use doubled and stitched nylon web halters with solid brass hardware, and ¾-inch cotton lead ropes with a finished length of 10 feet. The lead should have a quality snap spliced into one end.

Horses are generally difficult to catch because of fear, resentment, or habit. Unhandled horses will move away from a human because of the unfamiliar sight, sound, and smell. Other horses may evade you because they resent the treatment they get after they are caught: excessively hard work, ill-fitting gear, painful doctoring, or tactless training.

Whatever the initial reason, once a horse has successfully avoided being caught several times he may continue the behavior merely out of habit. Even though there may no longer be an unpleasant condition to blame for the action, the horse still shuns you. In some cases, avoiding being caught is just an enjoyable game for the horse.

The "walk down" method. Few methods produce as long-lasting results as the "walk down method" of catching. Because it is based on persistence, it may be time consuming at first, but it is worth the investment. Begin the lessons in a small enclosure

and gradually increase the size of the area.

When approaching the horse, don't be threatening in body language. Always walk toward the horse's shoulder, never his rump or his head. Move at a smooth, slow walk. Don't look directly at the horse, but keep him in your peripheral vision. When the horse stands and allows you to approach, scratch him on the withers and then walk away. Always be the first to turn and leave. Don't catch and halter the horse each time he lets you touch him. Soon it will take just a few seconds for the horse to decide to stand still.

Haltering in a safe, organized fashion prevents mishaps and bad habits. Approach the horse from the near (left) side and hold the unbuckled halter and rope in your left hand. With your right hand, scratch the horse on the withers and then move your right hand across the top of the neck to the right side. Pass the end of the lead rope under the horse's neck to your right hand and make a loop around the horse's throatlatch. The loop is held with your right hand. If the horse tries to move away at this stage, you can effectively pull the horse's head toward you while levering your right elbow into the middle of the horse's neck.

To halter, hand the halter strap with the holes in it under the horse's neck to your right hand which is holding the loop of rope. With your left hand, position the noseband of the halter on the horse's face and then bring the hands together to buckle the halter behind the horse's ears at the poll.

Turning loose. Turning a horse loose follows the same procedure in a somewhat reverse order. First, the loop is applied around the horse's neck, the halter is removed, and then the loop is released. The handler should hold the horse momentarily with the loop, then release the loop and gently push the horse away with the right arm or hand. If you are dealing with a chronic bolter, you might try dropping a few feed wafers on the ground before you turn the horse loose. He

SAMPLE MANAGEMENT CALENDAR

MONTH	FEED	VET/FARRIER	EXERCISE
JANUARY	Increase hay 10% for every 10 degrees below freezing	Farrier appt.	Out during day
FEBRUARY	Same	Deworm	Same
MARCH	Same	Farrier appt.; Beware of excess moisture on hooves	Same
APRIL	Gradually increase grain as work increases and/or gradually out on pasture.	Same as above; deworm; Flu, WEE/EEE, Tetanus shots; float teeth.	Same
MAY	Increase grain; Decrease hay	Farrier appt.	Same
JUNE	Rotate grazing of pasture as needed	Deworm including bots	Out at night
JULY		Farrier appt.; remove bot eggs	Same
AUGUST		Deworm/bot eggs	Same
SEPTEMBER	Organize winter water plan, if necessary	Farrier appt.; remove bot eggs; Influenza & Rhino booster	Out during day
OCTOBER	Decrease grain as work decreases	Deworm including bots	Same
NOVEMBER	Increase hay 10% for every 10 degrees below freezing	Farrier appt.	Same
DECEMBER	Same	Deworm	Same

soon will think more about inspecting the ground where he stands than about running away. As with any aspect of horse training, if you have problems handling your horses, seek the help of a qualified professional horse trainer.

Seasonal Considerations

Spring

Since spring is usually the wettest season, pastures have great potential to be damaged. It is also the time when pastures need to get a strong growth start, so it is best to confine the horses during the wettest periods. This will also keep their hooves in better condition as they will not be subjected to mud.

Poisonous weeds are some of the first plants to emerge so this is a good time to implement your eradication program.

Because horses are not used to the new green grass, introduce them gradually to the pasture. After they have had a full feed of hay, turn them out for a half an hour at a time, gradually increasing. Be wary of horses that tend to gain weight rapidly, get cresty necks, or are gluttonous, as they may be candidates for grass founder. Be sure the pasture is at least 4 to 6 inches tall before it is grazed, and remove horses when 50 percent of the forages have been removed.

If you have not been riding all winter, begin your riding program gradually. Be considerate of your horse's condition by comparing it to your own.

Summer

Be sure to make the time to ride often — this is what you have your horses for! Protect your horses from the sun by providing shade or by keeping your horse in during the day and turning him out at night. Keep flies

GROOMING	OTHER
Clean hooves daily all year	Turn compost as needed all year
	Spread manure
Shedding	
Check for ticks	
Clean sheath or udder	
	Buy Hay
Winter tail care	Spread manure
Body or trace-clip and blanket if working through winter	

under control and offer additional relief for pasture horses with the use of fly collars, tags, masks, and long tails. Keep an eye on your horse's weight — he can get fat and soft in a hurry on lush pasture. A horse's water and salt requirements increase with hot weather and exercise, so be sure your horses have plenty of both at all times.

Fall

Horses with decreasing activity levels should be receiving proportionately less grain. Supplement pasture horses with hay as needed so that the grazing season ends with some vegetation left in the pasture. Be sure water holes and creeks are still supplying ample fresh water for horses. Remove bots eggs and check for ticks.

Allow your horse to prepare for the stresses of winter well before the first snow. In September, horses living in temperate climates should be allowed an increase in

body weight of about 5 percent, but not more than 10 percent. A 1,200-pound adult should gain 60 to 120 pounds in the early fall. This extra flesh and fat will provide added insulation and an energy and heat reserve when weather is particularly bad.

Most horses begin shedding their summer hair in August and start growing thicker winter coats. A horse's normal winter coat has as much insulating capacity as most top-of-the-line blankets. The downward growth of the hair, coupled with the stepped-up production of body oils, sheds water and keeps moisture away from the skin. A dry horse has a much better chance of remaining a healthy horse. In order for your horse to produce a dense, healthy coat, his diet should provide an adequate quantity and quality of protein.

Winter

Winter-specific stresses include but are not limited to wind, wet, cold, lack of exercise, and owner disinterest. Horses can withstand temperatures well below freezing, as long as it is sunny and the air is still. The winter coat traps body heat next to the horse's skin. During cold temperatures, pilo erector muscles make the hair stand up, increasing the coat's insulating potential. Wind, however, separates the hairs, thereby breaking the heat seal and resulting in a great loss of body warmth.

Snow showers, sleet, and the freeze-and-thaw typical of some geographical areas are particularly hard on horses. A wet horse loses body heat many times faster than a dry horse does. In addition, wet hair tends to become plastered close to the horse's body, eliminating the air insulation layer.

A fuzzy winter coat can be deceiving if you are appraising your horse's condition by visual inspection alone. Feel the rib area for its flesh covering at least once every two weeks throughout the winter to monitor a horse's weight.

If a horse has a natural winter coat,

A horse "turned out" for the winter grows a long, fuzzy coat that stands erect and insulates the horse from the cold.

grooming just consists of a minimal "dusting" of the hair ends, or no grooming at all. Bathing, vacuuming, or vigorously currying a winter coat can disrupt the natural protective layer of oils which is essential for protection from moisture. After riding, rub the coat dry with a cloth or gunny sack or allow the horse to roll in clean sawdust, sand, or dry snow.

Ensure adequate water intake by checking a horse's water source twice daily. Horses can last for only three days without water. During the winter, they typically drink between 4 and 12 gallons a day. Although intake will be at the low end of the range during the cold season, the effects of dehydration can easily go unnoticed during winter months. Forcing horses to obtain their needed moisture by eating snow is counterproductive. In addition to the fact that six times as much snow must be eaten to provide an equivalent amount of water, horses must use precious body heat to melt the snow. This requires them to use up calories that could be used for warmth just to satisfy their thirst.

When the average temperature is or will be below freezing, increase your horse's roughage. For every ten degrees Fahrenheit

below freezing, the hay ration should be increased by 10 percent. When it is twelve degrees Fahrenheit (twenty degrees below freezing), the grass-alfalfa hay ration of a 1,200-pound horse may be increased from 24 pounds per day (using the usual recommendation of about 2 percent of the body weight as a base) to 28.8 pounds per day (a 20 percent increase: 24 pounds x 20 percent = 4.8 additional pounds, added to the initial 24 pounds equals 28.8 pounds). Horses fed less than is necessary to combat cold and wind will shiver to keep warm, burning fat and muscle tissue and thereby losing weight.

Contrary to popular belief, feeding grain will not appreciably increase a horse's body warmth, but feeding increased roughage will. The heat of digestion (in terms of calories) is greater and lasts longer from hay than from concentrates. It is most beneficial to feed a horse several hours in advance of a storm rather than during it. Immediately after a large meal, blood is concentrated in and around the digestive tract rather than in the muscles where it is needed for warmth.

Because of winter's snows and sunny thaws, feeds can spoil easily. Damp hay, pellets, or grain can become fermented or moldy in just a few hours in the sun. Check feed over carefully during daylight hours, then offer an amount that will be cleaned up in one feeding, removing what is left. Feeding whole bales for human convenience is wasteful and, if the feed spoils, could cause digestive problems. In addition, the twines are dangerous if a horse ingests them or gets tangled in them.

Slippery footing can cause a horse to work cautiously, at best, or cause it to lose its balance and slip or fall. Wrenches and sprains are common in winter, in addition to sole bruises from the lumpy frozen ground. Whether to shoe or to allow the horse to be barefoot for the winter is largely a matter of personal preference and is dependent on the condition of the horse's hooves and the use of the horse during the winter months. Confer with your farrier.

🐎 CHAPTER SEVENTEEN

Record Keeping

G OOD RECORDS ARE ESSENTIAL FOR FINANCIAL SUB-stantiation and for documentation of health care and other management activities. In order for records to be useful, they must be complete and accurate. Your record-keeping system should be designed with your personal habits in mind. If you make your forms easy to use and locate them in convenient places, you are more likely to make entries on a regular basis. Your record forms should be simple, but this does not mean using a blank sheet of paper to record everything about a particular horse! Categorize and organize the types of information you wish to record.

Management Records

Whether your horse venture is a hobby or business, you should keep accurate, complete records on health care, farrier work, training, and breeding.

Jotting everything on one calendar just doesn't work. Using a calendar to remind you of upcoming appointments and events is fine, but you will also need a place to record the various details related to horse ownership and care. A 5-by-8 card file works well. You can make dividers for each horse

and subdivide these further into categories (vet, training, farrier, etc.).

In the event of an emergency, you should always have a current set of feeding instructions that are easy for anyone to understand. A blackboard in the feed room works well for this.

In addition to daily records, you should keep a file folder on each horse to store documents such as a breeding certificate or bill of sale, registration papers, brand inspection, tattoo, freeze brands, permanent identification records (including photos and written description indicating height, weight, scars, brands), pedigree, and insurance policies.

Insurance Records

For your homeowner's or business property insurance policy, you will need to have an itemized inventory of your tack and equipment. List the major items with a brief description and serial or registration numbers if applicable. List the value and include receipts if possible. Photographs or video-tapes of your tack room inventory can also assist with identification and description. Keep one copy of the inventory and visual supports with your insurance policy at home and one in your safety deposit box.

Your homeowner's policy will usually cover public liability and property damage if the horse use is "recreational." But be sure you understand what your particular policy means by the difference between "recreational" horse use and a horse business. A business is indicated when you file a Schedule F (Farm) with the IRS. Often, however, even if you consider yourself a hobby farmer you may be treated as if you were running a horse business. If one of your horses is used for showing, parades, or rental, or if your facilities are used for boarding, training, instructing, or breeding, you may have to obtain additional liability coverage with a supplemental business policy.

Financial Records

If you are pursuing your horse endeavor as a business with profit intent, you should work closely with a tax counselor who has experience with farm accounts. You will have to keep formal income and expense records. You will need to keep your horse operation transactions separate from monies used for domestic matters, other farm ventures, hobbies, and your other work. One of the best ways to do this is to maintain a separate checking account for your horse venture. All equine expenditures and income and no others should be handled through this account. You can, of course, make cash withdrawals (hopefully profits) from the business account to use elsewhere.

Your horse business will need to show a profit intent in order to retain your business (farm) status. If that status is in question, the IRS will use specific guidelines to determine if your horse operation is designed with profit intent. They will look at the manner in which you carry on your activity and the type of bank accounts, records, and logs that you maintain. The IRS accepts the following types of documentation in an audit, in the order in which they are listed: receipts,

checks, logs, third-person verification, and verbal description.

They will note if you have sought advice from experts, not only horse-related advisors, but marketing and financial counselors as well. The IRS will evaluate the time and effort you expend in your business: is it a week-end diversion, a full-time business, or something in-between? Because of the new "passive activity" laws, it would be advisable to keep a log of the hours you actually work in your business. You may be asked verbally or in writing if you have expectation and basis that your assets will appreciate. You should have a projected profit plan available, which is updated each year.

The IRS will also review your success in both similar and different types of businesses you have had. They will look at the previous history of profit or loss of your current horse business. They may look at your overall financial status and see how the horse business relates to your total income picture. They will do this to try and determine if you are legitimately profit-motivated or if you entered the horse business to use the deductions to offset your income from another area. Also, because many non-horse people see only the pleasurable aspects of horse ownership, you may be asked to justify the business nature of your horse operation. You may know of the hard physical work, the early hours, and the many personal sacrifices that are required to run a successful horse operation, but an IRS representative may only envision you galloping across a meadow heading for a tax shelter.

If you are legitimately in the horse business, you have nothing to worry about. Just keep good records and find an experienced farm tax counselor. Your counselor will help you define and understand income and expenses in current terms.

Income

You must report all income related to your horse business. This includes horses bought for immediate resale, horses raised and sold, and assets acquired for the business and then sold (example: trailer, saddle, breeding stock). It also includes prizes and awards, show and track winnings, boarding fees, training fees, lesson fees, proceeds from hosting events and clinics, and income from the sale of farm-raised grain, hay, and straw.

If you engage in bartering, the services or goods you receive are considered income and must be recorded at the fair market value of the item or service. For your half of the trade in bartering, what you gave will be considered an expense if it would have been a deduction in the normal course of business.

Expenses

Horse business expenses generally fall into one of three categories: those operating expenses that are entirely horse related and fully deducted in the year the expense is incurred; those that are shared expenses with domestic or other businesses and must be prorated and then the horse portion deducted in the year incurred; and those major expenses that are spread out (depreciated) over several years.

The operating expenses are fairly straightforward. Either your expenditures are "ordinary and necessary" for your business or they are not. Be sure you fully understand the definitions and requirements for an expense to qualify for a particular category. Some expenses are only partially deductible (such as business meals) and others are only deductible if your horse operation is showing a profit (such as an office in the home).

Expenses that need to be prorated include vehicles, taxes, and utilities. In each situation, you must determine, to the best of your ability, what percentage of the expense is directly attributable to the horse operation. If a vehicle is shared between domestic and farm purposes, keep a log book with the starting and ending mileage each time you use the vehicle for business-related activities,

along with the business purpose of the trip. Record enough information in your log so that you can answer the following questions about your trip if asked by the IRS: Who? What? When? Where? Why? How? You may find it more expedient to keep the mileage notations on domestic activities instead, if they make up the smaller percentage of the total mileage.

Determine what percentage the business miles are of the total miles traveled by that vehicle during the year. Then your tax counselor will determine which is to your advantage — using a mileage allowance deduction (so many cents per mile) based on actual business miles traveled, or taking a percentage of actual expenses (total of gas, oil, repairs, etc.) based on the percentage of business use. Either way, for your tax counselor to make this assessment, you will need to keep all receipts concerning vehicle operation and maintenance as well as a mileage log to determine the business percentage use. If you are using a leased vehicle, you are only allowed to deduct actual expenses.

Property taxes are usually determined by consulting the itemized breakdown on your property valuation and tax statement. The tax resulting from the assessed value of horse-related buildings is deductible, as is the tax assigned to the portion of the land you use for the horse operation. If you have ten acres, you may be using two for house, yard, and other domestic uses and eight for the horse operation. The tax related to the eight acres is deductible.

Prorated expenses that may be a little tougher to divide up are utilities. With electricity, for example, unless you have separate meters for your barn and your house, you will have to estimate what percentage is farm related. If you are just adding horse facilities to your residential acreage, you can compare utility bills before and after the additions to help you calculate business-related kilowatt hours and dollars. To further help you determine your utility business percentage, you may be able to get hourly rate estimates

FARMERS' TAX SAVINGS INVENTORY

INCOME

Animals Bought for Resale

Name	Date Bought	Date Sold	Purchase Price	Selling Price
_____	_____	_____	_____	_____
_____	_____	_____	_____	_____
_____	_____	_____	_____	_____

Raised Animals Sold

Name	Birthdate	Date Sold	Purchase Price	Selling Price
_____	_____	_____	_____	_____
_____	_____	_____	_____	_____
_____	_____	_____	_____	_____

Assets Sold

Name	Date Bought	Date Sold	Purchase Price	Selling Price
_____	_____	_____	_____	_____
_____	_____	_____	_____	_____
_____	_____	_____	_____	_____

Prizes & Awards $ _____   Training Fees $ _____   Boarding Fees $ _____

Hay & Straw $ _____   Grains $ _____   Track Earnings $ _____

Barter $ _____   Other $ _____   Other $ _____

QUESTIONS: _____

_____

_____

DEDUCTIONS

Advertising $ _____   Bank Charges $ _____   Repairs $ _____

Rent — Pasture $ _____   Feed $ _____   Seeds/Plants $ _____

Machine Hire $ _____   Supplies $ _____   Breeding Fees $ _____

Vet & Med $ _____   Entry Fees $ _____   Training Fees $ _____

Travel $ _____   Legal & Prof $ _____   Licenses $ _____

Casual Labor $ _____   Commissions $ _____   Dues & Subscript $ _____

PRORATED DEDUCTIONS

Total Vehicle Mileage _____   Business Miles _____   Percentage Business Use _____

Total Gas & Oil _____   Business Total _____

Interest Total _____   Business Total _____

(How arrived business percentage) _____

_____

Taxes Total _____   Business Total _____

(How arrived business percentage) _____

_____

Utilities Total _____   Business Total _____

(How arrived business percentage) _____

_____

Assets purchased or converted to business use

Item	Date Bought or Converted	Purchase Price/FMV*

USE ADDITIONAL SHEETS WHERE NECESSARY
*FAIR MARKET VALUE

from your utility company on the kilowatts required to operate various electrical devices. Note that what you determine as a percentage for electricity may be totally different than the business percentage for gas, water, or other utilities. Most small horse operations located on a residence use an average of 20 to 30 percent of the total utilities.

Large assets purchased for the business (truck, trailer, buildings, tractor, etc.) or those converted to business use are often depreciated. Your tax counselor will either advise you to take a full expense in the year incurred for an item (depending on the cost of the item and the status of your profit or loss) or will determine what depreciation schedule each item will follow. Then each year of the depreciation schedule, a certain percentage of the cost of the tractor, for example, will be available to use as a deductible expense. It must be remembered, however, that during the year you sell a business asset, its expense potential will have to be recaptured and treated as income.

Whenever you pay $600 or more to an individual or a non-incorporated business in one year for goods or services, you are required by law to fill out a Form 1099 and send it to the IRS and to the person or business for them to report as income. So if your bills for veterinary work totaled $600 or more, you should issue a 1099 to your veterinarian. The same goes for your farrier, your hay supplier, etc.

This work sheet outlines some of the most common categories of income and expense related to a horse operation. Definitions and requirements for your expenditures to qualify for tax-deductible business expenses vary from year to year. Work closely with your tax counselor for a current interpretation.

SECTION FOUR
APPENDIX

 APPENDIX I

Space Requirements

··

SPACE REQUIREMENTS FOR HORSES IN BUILDINGS

ANIMAL	ANIMAL SIZE	BOX STALL SIZE	TIE STALL SIZE[1]
MATURE ANIMAL (Mare or Gelding)	small	10' x 10'	
	medium	10' x 12'	5' x 9'
	large	12' x 12'	5' x 12'
BROOD MARE		12' x 12'	
BROOD MARE WITH FOAL		12' x 24'	
FOAL TO 2-YEAR-OLD	average	10' x 10'	4½' x 9'
	large	12' x 12'	5' x 9'
STALLION[2]		14' x 14'	
PONY	average	9' x 9'	3' x 6'

[1]Including manger
[2]Work stallions daily or provide a 2-4-acre paddock for exercise

HAY MANGER AND GRAIN BOX DIMENSIONS

ANIMALS	HAY MANGER[1]	DIMENSIONS[2]	GRAIN BOX
ALL MATURE ANIMALS (Mares, Geldings, Brood Mares, Stallions)	30"-36"	Length	20"-24"
	38"-42"	Throat Height	38"-42"
	20"-24"	Width	12"-16"
	24"-30"	Depth	8"-12"
FOALS AND 2-YEAR OLDS	24"-30"	Length	16"-20"
	32"-36"	Throat Height	32"-36"
	16"-20"	Width	10"-16"
	20"-24"	Depth	6"-8"
PONIES	24"	Length	18"
	32"	Depth	32"
	18"	Throat Height	10"
	20"	Width	6"-8"

[1]Wall corner hay racks are often used instead of mangers. [2]Five feet is the usual distance between the floor and bottom of the rack. Many horsemen feed hay on the stall floor in both box and tie stalls and use a wall-mounted grain box in the corner of the stall.

STORAGE SPACE REQUIREMENTS FOR FEED AND BEDDING

GRAINS	LBS/BU	LBS/CU FT
CORN Shelled	56	44.8
CORN Ear	70	28.0
BARLEY	48	38.4
OATS	32	25.6
SOYBEANS	60	48.0

HAY/STRAW	CU FT/TON	LBS/CU FT
ALFALFA Loose	450-500	4.4-4
NON-LEGUME Loose	450-600	4.4-3.3
STRAW Loose	670-1000	3-2
ALFALFA Baled	200-330	10-6
NON-LEGUME Baled	250-330	8-6
STRAW Baled	400-500	5-4

 APPENDIX II

Resources

....

Write or call for more information and to find the dealer nearest you.

Barns and Stalls

Lester/Butler Rural Systems
P.O. Box 37
Lester Prairie, MN 55354
800-826-4439

Loddon Livestock Equipment
5280 S. University Blvd.
Greenwood Village, CO
80121
800-348-5982

Port-A-Stall Corp.
P.O. Box 1627
Mesa, AZ 85211
602-834-8812

Rohn Agri Products
P.O. Box 2000
Peoria, IL 61656
800-447-2264

Woodstar Products, Inc.
P.O. Box 444
Delavan, WI 53115
414-728-8460

Fence

Bayco Fencing
2060 East Indiana Ave.
Southern Pines, NC 28387
800-822-5426
(POLYMER FILAMENT)

Country Estate Fence
P.O. Box 45
Cozad, NE 69130
800-445-2887
(POLYVINYL)

Keystone Steel and Wire Co.
7000 S.W. Adams St.
Peoria, IL 61641
800-447-6444
(WIRE AND WIRE MESH)

Ranch Life Plastics, Inc.
4125 W. Harper Rd.
Mason, MI 48854
800-551-4348
(POLYMER AND POLYMER
WITH WIRE)

Spur Fence
2720 East Avalon Ave.
Muscle Shoals, AL 35661
800-348-7787
(POLYMER RAILS)

Stockyards Ranch Supply
6990 Vasquez Blvd.
Commerce City, CO
80022
800-443-5022

Tensar Polytechnologies
1210 Citizens Parkway
Morrow, GA 30260
800-845-4453
(POLYMER MESH)

Triple Crown Fence
P.O. Box 2000
Milford, IN 46542-2000
800-365-3625
(POLYVINYLCHLORIDE)

Horse Waterers & Feeders

Brower Equipment
P.O. Box 2000
Houghton, IA 52631
319-469-4141

Country Manufacturing, Inc.
P.O. Box 104
Fredericktown, OH 43019
614-694-9926

Farnam Equipment Co.
301 W. Osborn
P.O. Box 34820
Phoenix, AZ 85067
602-285-1660

Franklin Equiment
P.O. Box 271
Monticello, IA 52310
319-465-3561

High Country Technology
1175 Boeing St.
Boise, ID 83705
800-388-3617

Nelson Manufacturing Co.
P.O. Box 636
3049 12th St., S.W.
Cedar Rapids, IA 52406
319-363-2607

Ritchie Industries, Inc.
P.O. Box 730
Conrad, IA 50621
800-747-0222

Rubbermaid Commercial
Products
3124 Valley Ave.
Winchester, VA 22601
800-347-9800

Pipe Panels

HiQual
3139 Creek Dr.
Rapid City, SD 57701
605-343-1234

John Lyons
P.O. Box 479
Parachute, CO 81635
303-285-9797

Powder River
P.O. Box 50758
Provo, UT 84605
800-453-5318

Preifert Manufacturing
P.O. Box 1540
Mt. Pleasant, TX 75455-
1540
800-527-8616

Stall Accessories

Colorado Kiwi Co.
P.O. Box 808
Clark, CO 80428
800-345-8846
(LATCHES, HINGES)

Paul's Harness Shop
4255 Sinton Rd.
Colorado Springs, CO
80907
800-736-7285
(MISCELLANEOUS STALL
ACCESSORIES)

K&D Plastics Ranch Products
Highway 82 W.
Gainesville, TX 76240
800-635-7919
(BUCKETS)

BMB Animal Apparel
Manufacturers
1935 Walker
Wichita, KA 67213
800-842-5837
(HALTERS, BLANKETS, MISC.)

Stall Mats and Flooring

Equine Comfort
493 Route 22, Suite A
Pawling, NY 12564
800-MATTING

Equistall
1801-A Willis Rd.
Richmond, VA 23237
800-448-3636

Groundmaster Products
15101 Algoma Ave. NE
Cedar Springs, MI 49319
800-968-2930

Linear Rubber Products, Inc.
5525 19th Ave.
Kenosha, WI 53140
800-558-4040

Miscellaneous

Cease Fire
510355 Capital Ave.
Noak Park, MI 48237
800-338-9010
(AUTOMATIC FIRE EXTINGUISHING
EQUIPMENT)

Dyco Associates, Inc.
830 Hawthorne Lane
West Chicago, IL 60185
708-231-7000
(ANTI WOOD CHEW)

Interland Group, Inc.
Footings Unlimited Division
P.O. Box 235
Gray Summit, MO 63039
800-972-7251
(ARENA FOOTING)

Jacobsen Division of Textron
1721 Packard Ave.
Racine, WI 53403
414-635-1251
(FARM EQUIPMENT)

Ron's Farm Equipment
906 N. US Highway 287
Fort Collins, CO 80524
970-221-5296
(FARM EQUIPMENT)

Spalding Laboratories
760 Printz Rd.
Arroyo Grande, CA 93420
800-845-2847
(FLY PREDATORS)

Westwood Co.
50 Westwood Lane
Trout Creek, MT 59874
406-827-4675
(CHUTES/STOCKS)

 APPENDIX III

Bibliography

Ambrosiano, Nancy W., and Harcourt, Mary F. *Horse Barns Big and Small.* Ossining, NY: Breakthrough, 1989.

American Association for Agricultural Engineering and Vocational Agriculture. *Planning Fences.* Athens, GA: AAAE & VA, 1980.

Boyd, James S. *Buildings for Small Acreages.* Danville, IL: Interstate, 1978.

Boyd, James S. *Practical Farm Buildings.* 3rd ed. Danville, IL: Interstate, 1993.

Building Systems Institute Staff. *Metal Building Systems.* 2nd ed. Cleveland, OH: 1990.

Burch, Monte. *How to Build Small Barns & Outbuildings.* Pownal, VT: Garden Way Publishing, 1992.

Damerow, Gail. *Fences for Pasture and Garden.* Pownal, VT: Garden Way Publishing, 1992.

Davis, Thomas A. *Horse Owners' and Breeders' Tax Manual.* Washington, DC: American Horse Council, Inc.

Equine Research Staff. *Breeding Management and Foal Development.* Grand Prairie, TX: Equine Research, Inc., 1982.

Heldmann, Carl. *Be Your Own House Contractor.* Pownal, VT: Storey Publishing, 1995.

Hill, Cherry. *The Formative Years: Raising and Training the Young Horse from Birth to Two Years.* Ossining, NY: Breakthrough, 1988.

Hill, Cherry. *Horse for Sale.* NY: Macmillan, 1995.

Hill, Cherry, and Klimesh, Richard CJF. *Maximum Hoof Power.* NY: Macmillan, 1994.

Hill, Cherry. *Your Pony, Your Horse.* Pownal, VT: Storey Publishing, 1995.

Lawrence, Mike. *Step-by-Step Outdoor Stonework.* Pownal, VT: Storey Publishing, 1994.

Lewis, Lon. *Feeding & Care of the Horse.* Philadelphia: Williams & Wilkins, 1995.

Midwest Plan Service. *Horse Handbook: Housing and Equipment.* Ames, IA: Iowa State University, 1971.

O'Keefe, John M. *Water-Conserving Gardens & Landscapes.* Pownal, VT: Storey Publishing, 1992.

United States Department of Agriculture. *Plants Poisonous to Livestock in the Western States, Agriculture Information Bulletin 415.* Washington, DC: US Government Printing Office.

Whitehead, Jeffrey. *The Hedge Book.* Pownal, VT: Garden Way Publishing, 1991.

Winter, Tony A., LLM, and Marder, Sue, LLM. *Tax Planning and Preparation for Horse Owners* and *Bookkeeper for Horse Owners.* Ossining, NY. Annual.

Yost, Harry. *Home Insulation.* Pownal, VT: Garden Way Publishing, 1991.

Glossary

....

BODY CLIP: a partial or full clipping of winter hair at various intervals during the fall and winter to facilitate cooling out a horse in regular work. A clipped horse must be blanketed and have access to protected facilities.

BODY ROLLER: a band that encircles the horse's barrel over his stable blanket. The top of the roller (located on the horse's back) often has padded leather or metal nubs to discourage a horse from rolling over in his stall. The roller holds the blanket in place as well as helping to prevent a horse from being cast by discouraging rolling.

CAPILLARY REFILL TIME (CRT): an indicator of circulatory function or impairment. The time it takes for the blood (pink color) to return to the gums after a spot has been pressed with the thumb. Should be relatively instantaneous.

CAPPED HOCKS: hocks that have been damaged by injury, conformation, and/or work and now are thickened, especially over the point of the hock.

CAST: when a lateral recumbent horse's legs become wedged between his body and a stall wall making it impossible for him to get up.

Usually happens during rolling. A horse that remains cast, unnoticed for a period of time, may suffer colic or other intestinal or circulatory damage.

COLIC: digestive discomfort in the horse. There are several types and different treatments for each: impaction, excess gas, twisted intestine, damaged blood vessels to the intestine, etc. These may have been caused by dehydration, ingestion of sand, ingestion of too much or poor feed, cold water too soon after a work-out, etc. The horse that has colic is restless, uneasy, paws, wants to roll, and may break out in a sweat. Colic is one of the leading causes of death in horses.

CURB: a type of bit with shanks that works on leverage action; a lameness of the hock joint.

DRESSAGE: a style of training and riding that has European origins and is the only type of flat (not over fences) riding included in the Olympics.

DRILLED: when seeds are put down into the earth, rather than broadcast on top of it.

EQUINE INFECTIOUS ANEMIA: EIA, a disease

of the nervous system. The Coggins test determines whether a horse has been exposed to EIA or is a carrier. The disease is spread by mosquitoes and other biting insects. Currently there is no vaccine and no cure for EIA.

FETLOCK: the joint in the horse's leg where the cannon bone and the long pastern bone meet.

FOUNDER: see laminitis.

HANDS: a means of measuring horses. One hand equals four inches. Horses are measured from the ground to the highest point of the withers.

HEART GIRTH: the circumference of the horse's body just behind the front legs and over the withers. There is a strong correlation between the heart girth and the horse's weight.

HOOF HORN: the fingernail-like protein material that makes up hooves.

INFLUENZA: a respiratory disease of horses caused by a viral infection. Horses should be vaccinated once or twice annually for influenza as recommended by your veterinarian.

LAMINITIS: a debilitating condition of the hooves resulting from circulatory impairment initially caused by a metabolic imbalance from grain, rich feed, cold water after hard work, colic, work on hard ground, postfoaling complications, etc. Nutritionally caused laminitis seems more frequently to afflict overweight horses. Laminitis is very painful for the horse and may result in death or permanent unsoundness. The chronic case is called founder.

LEACHING: washing away, as in the leaching away of precious nutrients in the soil.

LONGE: to work a horse at all gaits on a long line in a circle of about 60 feet in diameter around the handler.

MORTISE: a notch, hole, or space in a piece of wood to receive a corresponding projection piece called a tenon.

PADDOCK: a grassy area of approximately ¼ acre to one acre that is designed to provide exercise as well as some nutrition.

PARASITES: Internal parasites of horses are commonly referred to as worms and include bloodworms (strongyles), roundworms (ascarids), pinworms (oxyuris equi), and bots (gasterophilus). Most internal parasites of horses spend a portion of their time migrating through the horse's tissues, causing destruction along the way. Parasite infestation can be greatly reduced by sanitation of facilities, proper manure management, and a regular deworming program every two months.

PASTURE: a grassy area usually larger than one acre that is primarily designed and maintained to provide nutrition.

PEN: a non-grassy area that is designed primarily for outdoor living quarters for horses.

PERENNIALS: those plants that have a life cycle of more than two years.

PINCH TEST: a means to determine a horse's level of dehydration. A tent of skin is lifted on the horse's neck and then let go to return to its original flat position. The resiliency of the skin indicates its water content. Skin that remains peaked indicates dehydration.

PULSE: the heart rate of the horse determined by feeling the blood pumping through the arteries or by listening to the heart with a stethoscope. Normal resting adult heart rate is approximately forty beats per minute.

RABIES: a disease of the central nervous system that can be transmitted to horses by the bite of an infected animal, such as a dog or skunk. It is characterized by choking, convulsions, and the inability to swallow liquids. It is fatal if not treated. A vaccine is available for horses and it should be administered on your veterinarian's recommendation in areas of high incidence of rabies.

RESPIRATION RATE: the number of times a horse inhales and exhales in one minute. Normal resting adult respiration is twelve to fifteen breaths per minute.

RESTRAINT: controlling a horse by psychological, manual, mechanical, or chemical means.

RHINOPNEUMONITIS: an upper respiratory tract infection of young horses that can also cause abortion in pregnant mares. There is a modified live vaccine to administer to young horses and a killed vaccine for pregnant mares.

RUN: a long pen designed to encourage a horse to exercise.

SACKING OUT: systematically introducing the horse to potentially frightening things so that he learns not to fear them and will not panic.

SLEEPING SICKNESS: encephalomyelitis. A disease with inflammation of the brain and spinal cord. There are two strains to vaccinate against routinely: Western and Eastern. In some areas, it may also be necessary to vaccinate against the Venezuelan strain.

SOFFIT: the underpart of an overhanging eave.

STABLE RUBBER: the name given to a rag used for finishing or setting the coat in grooming. Often made of linen.

STOCKS: a chute-like enclosure that restrains a horse for veterinary care.

STRANGLES: equine distemper caused by the bacteria Streptococcus equi. Highly infectious and characterized by inflammation of the respiratory mucous membranes. There is a vaccine for strangles.

SURCINGLE: a piece of training tack that encircles the horse's heart girth and acquaints him with pressure. It can also be used over a stable blanket to keep it from turning.

TAN BARK: any bark that has had the tannin extracted. It is used for arena footing.

TEMPERATURE: internal body heat. Normal resting adult temperature ranges from 99.5 to 100.5 degrees Fahrenheit in horses.

TETANUS: an infectious disease of the nervous system caused by a bacterial toxin and commonly associated with puncture wounds. Horses should be vaccinated against tetanus annually.

THRUSH: a disease of the hoof that can cause decomposition of the frog and other hoof structures; usually associated with unsanitary conditions.

TRACE CLIPPED: a type of partial body clip.

WINDROW: rows of cut and raked hay that are drying and waiting to be baled.

WISP: a woven straw or hay pad used in grooming.

WITHERS: the prominent area of the horse's spine where the neck and the back join.

Index

....

OTHER STOREY BOOKS WE THINK YOU WILL ENJOY

Becoming an Effective Rider: Developing Your Mind and Body for Balance and Unity, by Cherry Hill. Teaches riders how to evaluate their own skills, plan a work session, get maximum use out of lesson time, set and achieve goals, and protect themselves from injury. 192 pages. Paperback. ISBN 0-88266-688-6.

Fences for Pasture and Garden, by Gail Damerow. The complete guide to choosing, planning, and building today's best fences: wire, rail, electric, high-tension, temporary, woven, and snow. Also chapters on gates and trellises. 160 pages. Paperback. ISBN 0-88266-753-X. Hardcover. ISBN 0-88266-754-8.

Horse Handling & Grooming: A Step-by-Step Photographic Guide, by Cherry Hill. This user-friendly guide to essential skills includes feeding, haltering, tying, grooming, clipping, bathing, braiding, and blanketing. The wealth of practical advice offered is thorough enough for beginners, yet useful for experienced riders improving or expanding their skills. 160 pages. Paperback. ISBN 0-88266-956-7.

Horse Health Care: A Step-by-Step Photographic Guide, by Cherry Hill. A step-by-step handbook explaining bandaging, giving shots, examining teeth, deworming, and preventive care. Exercising and cooling down, hoof care, and tending wounds are depicted, along with taking a horse's temperature, and determining pulse and respiration rates. 160 pages. Paperback. ISBN 0-88266-955-9.

Horse Sense: A Complete Guide to Horse Selection & Care, by John J. Mettler, Jr., D.V.M. The basics on selecting, housing, fencing, and feeding a horse including information on immunizations, dental care, and breeding. 160 pages. Paperback. ISBN 0-88266-545-6.

Keeping Livestock Healthy: A Veterinary Guide to Horses, Cattle, Pigs, Goats & Sheep, by N. Bruce Haynes, D.V.M. Provides in-depth tips on how to prevent disease through good nutrition, proper housing, and appropriate care. Includes an overview of the dozens of diseases and the latest information on technologies livestock owners need to know. 352 pages. Paperback. ISBN 0-88266-884-6.

101 Arena Exercises: A Ringside Guide for Horse and Rider, by Cherry Hill. A one-of-a-kind ringside exercise book for riders who want to improve their own and their horse's skills. 224 pages. Paperback. ISBN 0-88266-316-X.

Safe Horse, Safe Rider: A Young Rider's Guide to Responsible Horsekeeping, by Jessie Haas. Beginning with understanding the horse and ending with competitions, includes chapters on horse body language, pastures, catching, and grooming. 160 pages. Paperback. ISBN 0-88266-700-9.

Your Pony, Your Horse: A Kid's Guide to Care and Enjoyment, by Cherry Hill. Designed to complement riding books for young readers, this book focuses on all aspects of horse husbandry as well as equine activities for kids, providing information on selecting a horse or pony, sheltering, grooming, feeding, and health problems. 160 pages. Paperback. ISBN 0-88266-908-7.

These books and other Storey books are available at your bookstore, farm store, garden center, or directly from Storey Publishing, Schoolhouse Road, Pownal, Vermont 05261, or by calling 1-800-441-5700. We invite you to visit us on the Internet at www.storey.com.